W9-AEO-400

In Praise of Plants

In Praise of Plants

Francis Hallé

translated and with a foreword by David Lee

TIMBER PRESS
Portland • London

All drawings by Francis Hallé

Copyright © 1999 as *Éloge de la Plante, pour une Nouvelle Biologie*
by Éditions du Seuil, Paris

Translation copyright © 2002 by Timber Press, Inc.
All rights reserved.

Published in 2002 by Timber Press, Inc.
The Haseltine Building   2 The Quadrant
133 S.W. Second Avenue, Suite 450   135 Salusbury Road
Portland, Oregon 97204   London NW6 6RJ
www.timberpress.com   www.timberpress.co.uk

Printed in the United States of America

ISBN 978-1-60469-262-4

The Library of Congress has cataloged the hardcover edition as follows:

Hallé, Francis
    [Éloge de la plante. English]
    In prase of plants / Francis Hallé; translated and with a foreword by
        David Lee.
        p.   cm.
    Includes bibliographical references (p. 305).
    ISBN 0-88192-550-0
    1. Botany. I. Title.

    QK45.2 .H325513   2002
    580—dc21   2002017988

for Gabrielle
Daya
Raphaël
Michaël
and Léa
who have slowed down my work
and helped me to see the poetry of life

Contents

Foreword

David Lee

In Praise of Plants is a remarkable book by Francis Hallé, a French tropical botanist of international renown. It is an appreciation of plants from his unique viewpoint, both scientific and emotional. It is the culmination of more than 40 years of the study of plants and life in societies that have a close relationship to plants. The book also exposes English-speaking readers to the sensitivity of French culture to plants, with its long history of horticulture and its literary traditions.

Francis's first experiences of plants and animals were as a toddler at his parents' cottage and garden outside Paris during the German occupation. His world was 2 hectares: half a hectare of vegetable garden, half a hectare of pasture with various domestic animals, then a forest woodlot of a hectare at the back of the property. To an infant the forest was a mysterious and compelling place. Even today, Francis appreciates that a garden of 5000 m^2 can nourish nine people, including five perpetually hungry adolescents. He also remembers the war: his family hiding in a shelter, bombs exploding, wounded neighbors, the sky full of the parachutes of the first Americans, and Patton's army, whose soldiers gave him peanuts and chewing gum.

Francis was strongly influenced by his mother's love of gardening. She taught him the names of many plants. As the youngest of seven children, he watched his brothers grow up and travel to the Tropics. Nicolas, 12 years older and a passionate student of natural history, started as a scientific illustrator, first drawing crabs in Senegal, then plants in the Ivory Coast. He eventually became an expert on the taxonomy of West African plants. At about the same time, Francis's brother Noël left as a soldier for the war in Indochina, and Grégoire

was planting bananas in Guadeloupe. While Francis was studying in
Paris he received a stream of letters and packages with seeds, butterflies, dried fishes, snake skins, and *Raphia* palm fruits. Francis became, in his word, a tropicalist, and a serious student of tropical biology. He also nurtured another interest, architecture, and spent his
holidays sketching medieval monuments. His interests in architecture and plants came together in a remarkable way later in his life.

Francis's first experience of the Tropics, the Ivory Coast in 1959,
was an inspiration to him. He took up doctoral studies there and
continued working in the Tropics as a scientist in French Guiana.
These long stays, along with visits to other areas in the Tropics,
gave Francis an excellent firsthand experience of tropical rain forest
plants. This knowledge, combined with his interest in architecture
and skills as an artist, combined in a major contribution to botany:
the concept of plant architecture together with a system for comparing growth models. His book, *Tropical Trees and Forests, an Architectural Analysis* (1978), coauthored with Roelof Oldeman and Barry
Tomlinson, remains one of the most frequently consulted in tropical botany. He moved back to France as a professor in the Botanical
Institute at the University of Montpellier, where he continues in a
very active retirement.

I first experienced nature and the profound differences between
plant and animals, explored by Francis Hallé in this book, in my
mother's garden at the edge of a small town on the Columbia Plateau
of Washington state. It was 1945, I was less than 3 years old, and I
saw the bright green spears of iris and gladiolus pushing up in the
dark soil. Then I saw the quick movements of a brilliant blue snake,
a western racer. I explored the desert outside my town, and later the
more lush coniferous forests of the Cascade Range. My vocation as
a tropical botanist emerged during a trip to the South Pacific in 1964.

I first met Francis in 1975 when I was a lecturer at the University
of Malaya in Kuala Lumpur. After completing my doctorate, I studied the biochemistry of plant tissue culture in a windowless building
at Ohio State University. Less than satisfied, I obtained a university
position in tropical Asia. My frequent forays into the tropical rain
forests of Malaysia were a revelation to me and raised the research
questions that I still pursue today. Francis had been appointed as
external examiner in our department, and we became fast friends
from our frequent investigations of the forest near my rural home.

Francis invited me to participate in his laboratory as a visiting professor, to assist in the training of students in a graduate diploma in tropical botany. Students came from throughout Europe and various tropical countries, and this program was an important contribution to the development of research in tropical botany in Europe. During the year together in Montpellier, and for a month in French Guiana, we discussed some of the ideas of this book. After my family's return to the United States, Francis accepted a post in Indonesia, living there 2 years with his family. This stay in Southeast Asia completed his familiarization with the biology of plants throughout the Tropics.

Frustrated at the difficulty of studying plants in the rain forest canopy, Francis had proposed constructing a hot-air balloon and raft that would fly over the rain forest and deposit scientists on top of the canopy—a radical idea. After returning to France he found a pilot and architect to help him, and the concept developed into the hot-air dirigible known as *le radeau des cimes* (Figure 67). The canopy raft has enabled teams of scientists to study tropical rain forest canopies in French Guiana, Cameroon, Gabon, and Madagascar.

Francis told me more about the present book during a symposium at the University of Malaya in 1998, commemorating the career of our friend Benjamin Stone. He also urged me to participate in the canopy expedition in Gabon that year. Eight months later at La Makande in Gabon, we discussed the book again, walking in the forest together, and I agreed to translate it. It was easy to find the publisher because of the enthusiastic support of Timber Press.

There are several themes in the book that are of particular interest to me. The first is our human bias toward animals and relative indifference to plants. As I translated the book I watched my 1-year-old grandson begin to acquire language skills. The first words he learned were names of animals: dog, cat, bird. Even though I showed him fragrant and colorful flowers, the words flower, plant, and tree only appeared in his vocabulary much later. Francis points out our animal biases throughout the book.

Our concern about the loss of biodiversity (or megadiversity in the sense of Myers et al. 2000) is particularly oriented toward animals. The symbols used in campaigns to preserve nature are animals: panda, quetzal, bonobo, poison arrow frog, even a Komodo dragon. In the ecosystem approach to conservation advocated by Myers et al., plants are generally in the background as the environment for animals. When we speak of the value of ecosystems as providing services (clean water and air, for example), these are almost exclusively the benefits that plants provide. Plants only receive a little attention when we worry about the loss of potential drugs.

Francis feels a profound connection with nature through plants. Paul Shepard (1996), who wrote about the human relationship to nature, described our fundamental relationships to animals with profundity yet struggled to describe a similar relationship to plants, although he reckoned that such a bond existed. There is evidence for a deep-seated human relationship with plants in the popularity of gardening and the demonstrated value of horticultural therapy and plant-dominated landscapes (Kellert and Wilson 1993). *In Praise of Plants* moves along that path of discovery.

French culture, particularly noticeable in its literary traditions, expresses such a relationship with plants and the landscapes formed by them. At times, Francis writes descriptively and poetically within this tradition. He also quotes liberally from the literature of his cultural tradition. That same culture has developed a language and vocabulary through the cultivation of plants; many of these terms translate rather poorly into English. I am reminded of the crude term we use for an opening in the forest: gap. The French word *chablis* has a very specific meaning, conveying the process of that opening. The quotations at the beginning of each chapter, mainly by French authors, set the tone for the ideas that follow. Thus the book not only reveals the ideas of an eminent scientist, but the sensibility of his French culture.

Another value of this book is its use of comparison to illuminate familiar subjects. We are animals and we know about them. Yet our appreciation for animals is heightened by their comparison with plants. Interest in plants is certainly increased by comparing them with animals. Much of Francis's and my experience of plants has been in the Tropics. Our appreciation of the plants of temperate regions is dramatically enhanced by that experience. *In Praise of Plants* definitely has a tropical perspective.

The intellectual heart of the book is its analysis of plants from a few simple principles, particularly in comparison with animals, to build a unique and coherent picture of plant biology. Plants produce enormous surfaces for capturing resources. This requires that they be sessile, in turn exposing them to rapid environmental changes. Their lack of mobility has led to a remarkable diversification in their biochemistry, partly to entice animals to do their work for them. The life cycles of plants combined with the lack of separation of soma and germ have resulted in fundamentally different patterns of inheritance and, ultimately, evolution.

This is a controversial book. Francis uses some strong language and even stronger graphics to make us confront the differences between plants and animals; I suppose that some readers may be offended. Much of it is also very humorous. (This scientist could well have ended up writing comics!) Every reader will find some point of disagreement; I did. More importantly, readers will have their understanding of plants fundamentally altered, and their appreciation immeasurably enhanced.

Most controversially, Francis is not content to put his feelings about plants, about nature, and the loss of diversity in one corner and write about the science of plants in another. His opinions and science are mixed on almost every page. Scholars and scientists in particular are wary of doing this for fear that their feelings will be seen as a loss of objectivity. For biologists who study organisms in the Tropics a feeling that accompanies their study is grief—for the awareness that what we study will most likely disappear through destruction. That grief, suppressed, can be personally debilitating (Windle 1992). I feel that grieving in this book, more than Francis's dissatisfaction that people do not give plants their due. Beyond that, Francis shares a profound affection, yes love, for the plants studied all these years, even in the face of indifference and impending loss. My intuition is that we have a profound emotional connection to the green mantle that supports and surrounds us, and that has preceded our own existence. We need to explore that connection more deeply, and that quest is central to our own existence as we look for meaning in our lives as individuals and as part of the natural world.

> The world is fast losing its soul
> But you don't have to surrender yours.
> You don't have to live on a mechanical globe.
> You don't have to tame your deep-forest passions.
> You don't have to suppress your radiant beauty.
>
> THOMAS MOORE as quoted by
> Roberts and Amidon (1999)

A recent memory of Francis Hallé is a day in March 1999, in the Forest of the Bees, Gabon, in the cool darkness of 5 A.M., with strong hot coffee and the inflation of the dirigible for its morning mission. Launch is at dawn with Francis behind the pilot, Dany Cleyet-Marrel, and next to Gilles Ebersolt, the architect and designer, all close friends. They return, we deflate and store the dirigible, take a late breakfast, and process samples for biochemical analyses (Downum et al. 2001). Then lunch, rest, and a long walk in the okoumé forest down the Makande Stream. Francis sees something unusual. We stop while he writes notes and adds a sketch to his field book, the habit of a lifetime of field research. We return for afternoon tea. Francis tends his small garden, full of the plants he will carry away alive at the end of the expedition. Then, pastis and camaraderie before dinner.

I THANK Marcel Welch, a colleague and professor of French literature, for assistance with many of the literary translations. Leon Cuervo, a biology colleague, translated the quotation from Spanish into English. My editor at Timber Press and I benefited from expert advice from colleagues: John Dawson, Jim Eckenwalder, Jim Hamrick, Don Kaplan, Lydia Kos, Phil Stoddard, Walter Goldberg, Dan Schwartz, and George Yatskievych. Carol, my wife, supported me when circumstances made it more difficult to complete this project. Above all, I thank my mother, Mary Lee of Ephrata, Washington, who introduced me to nature through gardening.

Preface

Francis Hallé

THERE ARE many reasons why animals inspire our interest—admiration, a desire to collect them, photograph them, raise them for the butcher before eating them, use them as means of transportation, as experimental material in medical research, or as companions in a lonely life. It is clear that most of us have a spontaneous interest in animals. In contrast, this same majority is generally poorly acquainted with plants, looking down on them or simply ignoring them. Since plants do not move or make sounds, many do not even consider them to be alive. The psychological roots of this asymmetry in our perception of the two kingdoms deserve analysis, the allures that animals and plants exercise on us being different.

As a result of this bias, which often borders on injustice, the idea has arisen in me to show how plants and animals differ yet resemble each other. Comparisons are made in morphology, growth and embryogenesis, cellular structure and function, biochemistry and communication, genetics and evolution, energy use, and ecology. Certain groups of organisms—fungi, corals, social insects—enrich the comparison. It appears that plants, often thought very inferior to animals, surpass them in a number of ways biologically.

If humans seem more easily interested in animals, if animals are attended to more spontaneously than plants, it is a result of zoocentrism or anthropocentrism. We are attracted by what we resemble, but we remain indifferent to that which is not cast in our own image. This mechanism of identification may provide an adaptational advantage. Humans, remarkably equipped to perceive movement, have applied this talent to hunting and warfare, but this mechanism also presents an obvious risk: self-absorption.

Biology has much to gain from ridding itself of such errors of perspective, which do not have a place in science. Besides, those who admire plants will find in the comparisons that fill this book additional reasons to feel pleased by the presence of plants, their closeness, and intimacy. What better antidote to the fearful constraints of urban existence than a garden planted with trees? Plants also have the merit of leading us directly to the roots of our primordial consciousness.

A few words of explanation of a technical nature are necessary. The terms animals and plants, without any additional explanation, designate animals endowed with active mobility—slug, termite, manta ray, great-crested grebe—and terrestrial plants fixed in place to a substrate, generally soil—fern, morning glory, daffodil, mango, *Araucaria* (Figure 1). More precise designations are used for algae, mosses, and colonial animals—sponges, corals—as well as for living creatures that are neither animals nor plants—bacteria, protists, mushrooms.

A general classification of cellular organisms (excluding viruses and prions) was published in a classic book by Lynn Margulis and Karlene Schwartz (1982) in which five kingdoms are distinguished:

> Prokaryotae
> Protoctista
> Fungi
> Animalia
> Plantae

I have modified this classification concerning the algae. They will not be considered as protists (the Protoctista of Margulis and Schwartz) but as plants, always a little peculiar because they do not have embryos or roots. [Profound differences among organisms within the protists have led to the recognition of an additional kingdom, the Archaebacteria, but such differences are not the subject of this book. —translator's comment] I continue to use the term kingdom despite, or perhaps because of, its humorous dimension. We can imagine a king of animals, but who would be the queen of plants or the monarch of mushrooms? Nevertheless, I see no reason to abandon a term that is both pleasant and venerable.

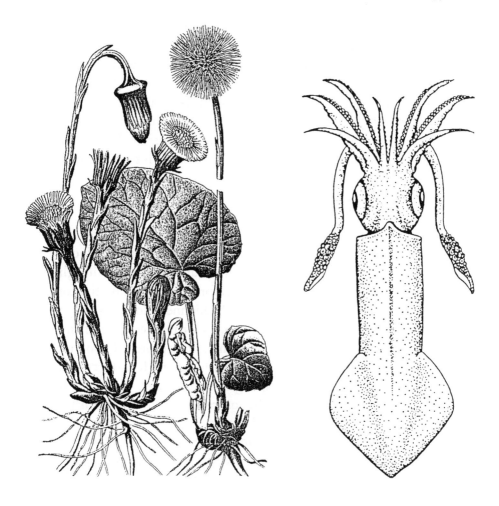

Figure 1. A plant and an animal: a coltsfoot, *Tussilago farfara* (after Masclef 1891), and a squid, *Loligo surinamensis* (after Biaggi and Arnaud 1995).

I would like to make several suggestions to the reader. If specialists read the book, they will find exceptions in their respective subjects to the general ideas to which *In Praise of Plants* is devoted. Such is the nature of diversity, and odd cases are part of the beauty of life. However, one of the dangers threatening the natural sciences is that peculiar cases spark such interest that we risk losing our view of the general laws of nature, whose manifestations are often less visible or more austere. I do not advise reading this book only with the intent of finding exceptions to the general rules that are presented. Criticism is obviously welcome, but I hope that the general concepts are conveyed without suffering from the tiny exceptions. Here and there the reader will find some strange images, crafted according to a simple principle: to impart a plant-like behavior to animals. Their aim is to shake off our habitual views, to make the profound differences between the two kingdoms more tangible. Is it not surprising, with each so beautiful, that their mixture is so grotesque?

Acknowledgments

I first direct my thanks to those who have helped with the topics on animals. I thank François Bonhomme for discussions of genomic function and evolutionary mechanisms, Bruno Corbara for thoughts on social insects, Rosine Chandebois for information on the constraints of the embryonic state, Jean Chaudonneret for ideas on symmetry and polarity, and Robert Connes for intellectual and emotional support from the very beginning of this project. I also thank Jean-Marc Dauget for ideas on coral architecture, Jean-Pierre Diaz for precious information on sponges, Louis Euzet, who made me understand the importance of the contrast between animals that are freely moving versus those fixed in place, Rémi Gantès for contributions on mollusks, Yves Gillon, whose zoological knowledge I have admired for more than 40 years, Nicolas Hallé, to whom I owe the beautiful examples of asymmetry, Jacques Laborel for contributions on plasticity in corals, Mark Moffet for comparisons of the two kingdoms, Georges Petavy for help on the embryology of insects, Maria Prost, who authored one of the rare attempts to compare plants and animals, André Raibaut for everything concerning the embryology of vertebrates, Brian R. Rosen, who continually assisted me with corals, and Vincent Savolainen, one of the first scientists to encourage me in this project and whose enthusiasm has inspired me so much. I also thank Yves Turquier for ideas on the roles of the three embryonic layers, J. E. N. Veron, whose book on corals has been a source of inspiration, Mylène Weill for valuable help in immunology, and Michel Pichon and Helmut Zibrowius for everything concerning coral biology.

Despite my familiarity with plants, it is on this subject that I have needed help from friends and colleagues. I thank Paul Barnole for his unified concept of photosynthesis, Daniel Barthélémy for the connection between virology and geographic convergence, Richard Batemen, who conceived of the tree as a colony of genotypes and who introduced Marshal W. Darley to me, Adrian D. Bell for articles on bilateral plants and trifids, Yves Caraglio for help in the measurement of surfaces, Anne-Marie Catesson, to whom I owe the idea of lignin as excrement, Jacques Crabbé for ideas on biological types, Pierre Cruiziat for ideas on apoptosis, Paul Champagnat, who helped me appreciate the work of Anthony Trewavas, Henri Darmency for information on the adaptation of weeds to herbicides, Claude Edelin for the concept of architectural unity, Nicole Ferrière for her interest in this project from the beginning and for discussions on the levels of organization of cells and tissues. I thank Hubert de Foresta for sharing his insight on reiteration in termite colonies, Dawn Frame for her stimulating discussions on plant evolution, Jean Gaultier for assistance in paleobotany, Pierre-Henri Gouyon, to whom I owe much on the controversial subject of evolutionary mechanisms, Nicolas Hallé, from whom I take much of my love for botany, John L. Harper, whose work has always had an important place in comparisons of plants and animals, Rose Hébant for ideas on the function of meristems, Christophe Jourdan for contributions on the surfaces of roots, Francis Kahn for comparisons of embryology in the two kingdoms, Donald R. Kaplan, whose encouragement provided the necessary incentive to compare trees and corals, Roland Keller, who knew how to treat vegetative characters taxonomically, Michel Lauret for his knowledge of algology, Pierre Lieutaghi and Marianne Mesnil for assistance in ethnobotany, Jean Matricon for information on energy flux, Darlyne Murawski for research on genetic polymorphisms within a single tree, Philippe de Reffye for ideas on the importance of physical factors in plant form, and Jean-Claude Roland for his concern with cellular organization. I experienced all this scientific dialogue in a refuge of dispassionate courtesy, one of the great pleasures in doing research.

I give very warm thanks to Jeannine Blanc and Richard Palomino, who checked the citations and illustrations, and Yves Caraglio, Jean-Claude Combes, Rémi Gantès, Yves Gillon, Nicolas Hallé, Jean-Marc Lévy-Leblond, and Thierry Thomas, who helped improve the text by their critical reading, each in his or her specialty. If any errors have escaped notice, they are certainly my responsibility.

CHAPTER 1

Plants, Animals, and Humans

How treelike we are,
How human the tree
　　　　　GRETEL EHRLICH, "River history," 1998

"Where are the men?" the little prince asked, politely.
The flower had once seen a caravan passing.
"Men?" she echoed. "I think there are six or seven of them in existence. I saw them, several years ago. But one never knows where to find them. The wind blows them away. They have no roots, and that makes their life very difficult."
　　　　　ANTOINE DE SAINT-EXUPÉRY, *The Little Prince*, 1943

Plants know a lot about depths, and what goes on down there.
　　　　　PIERRE LIEUTAGHI, *La Plante Compagne*, 1998

HUMANS have always preferred animals to plants, no matter what the time and place. I state this with scarcely any fear of seeming demented. All the evidence suggests that animals fascinate us. They capture our attention. They arouse various feelings, of admiration, curiosity, lust, often sympathy, sometimes fear and repugnance, but never indifference. Age is not an issue in this fascination; even a young child knows that the stirring of oak leaves is caused by the wind or an animal. "Look! Look! There is an animal there," says my granddaughter Gabrielle, 4 years old. Urgently, Gabrielle seeks the identity of the little animal hidden in the foliage. Is it a magpie? A squirrel? Or only the neighbor's cat? Even before the animal has revealed itself, Gabrielle has forgotten the tree that holds it. The oak has become no more than a "plant landscape" (Brosse 1958). The tree is so enormous and immobile that Gabrielle is certain it will be there for a few more moments, or tomorrow.

It is significant that the child is spontaneously interested in animals but not in plants. Immobile and silent, what can they do to
seduce her, she who runs, cries, skips about, and sings from morning
to evening? Children are not alone in this preference; adults prefer
animals just as much. In the Tropics, loggers produce the tragic
scenes of destroyed forests, assisted by high technology. Directing
monstrous machines, they fell ancient trees with the same disinter-
est that they would kick an empty Coke can, inadvertently opening
deep wounds in the forest. They save their passion for hunting. If
the loggers see a troop of monkeys, they immediately stop their
machines, jump to the ground, sort out their machetes and rifles,
cut a trail, and move quickly and silently into the forest, agile and
capable, aroused, focused, fierce, on their way to battle.

I anticipate your objections: Certainly, these are landless peasants, victims of the rural exodus, scheming to feed their families.
Besides, is it not normal to observe such different values so far from
where we live? Well, such attitudes prevail even among our neighbors, revealing a preference for animals, completely and scornfully
ignoring plants.

Who Cares About Palms?

Replacing elephant ivory with vegetable ivory has been presented as
an environmental victory. However, who is concerned about the
bleak future for the palms that provide the vegetable ivory, the
strange *Phytelephas* of the understory in the western Amazonian rain
forest, whose existence is threatened by the destruction of the last
primary forests at the foot of the Andes? Would we prefer to see the
palms vanish rather than the elephants?

At the supermarket I found a shampoo, "Pure, natural, with essential oils of sage and juniper berry, chamomile, orange, and wood
rose." The label stated nicely, "This product was not tested on animals." Why do we have a Society for the Prevention of Cruelty to
Animals but not one for prevention of cruelty to plants? Why do we
have leagues against vivisection but do not protect plane trees
against pollarding by arborists?

This difference is not new, as Pierre Lieutaghi (1991) has noted:
"At Lascaux, already, nothing is seen of plants; the animal-king
seems to be the object of all their devotion. . . . Moreover, it is cer-

tainly plants that sustain the reindeer and the bison. And the hunter also eats fruits and grains. It is merely that plants do not offer any chance for glory, that they are not subdued dangerously. Nothing for the memories of adventurers. . . ." This is the attitude today. In 1990, EDF (Electricité de France) commissioned the damming of the Sinnamary River at Petit-Saut, French Guiana. In the forest soon to perish, scientists mobilized for the rescue of animals—monkeys, turtles, sloths, armadillos—all of whom could save themselves by swimming. Not a tree or liana benefited from this rescue effort. Plants cannot swim, and flooding dooms them. Noah's mistake is repeated. Petit-Saut? The greatest destruction of forest in the entire history of French Guiana.

In February 1997, a sheep was cloned for the first time, with plenty of media hype. Cloning plants, a horticultural technique and agricultural practice since time immemorial, has never particularly interested journalists. However, with sheep we are getting close to human beings.

Let us accept the evidence. Humans have a passion for animals even if they come to prefer them on the plate rather than admiring them for their freedom. Interest in animals persists as a deep emotion, completely accessible and common to all, profoundly natural. We cannot say the same about plants. Humans "are preoccupied less to know them than to understand them, less to understand them than to control them, and even more to control them in order to use them" (Brosse 1958). Plants to us are principally food, drink, medicine, raw material for industry, pasturage for domestic animals, green space for cities, landscapes for relaxation. They do not arouse any real passion in most of us.

The word botany, the name for the science of the study of plants, is derived from animals. In ancient Greek, *boton* was a herd animal, for which *botanê* was the fodder (Rey et al. 1992). The imbalance between our perceptions of animals and plants has very deep roots. *Animal* comes from *animate*, meaning capable of moving by itself, but the original meaning of that word, that which has a soul, also implies that plants lack souls. Such words appear harmless but often conceal powerful ideas. The imbalance is not only part of the European mind. Islamic tradition *(hadith)* allows images of plants but forbids those of animals and, *a fortiori*, those of humans, so as not to rival Allah. In traditional Islam, plants are not creatures of God.

Are Plants Alive?

To most people, plants are not even alive. At the youth center in a small town in Hérault, France, I explain that I am studying the differences between plants and animals. The kids tell me, "Oh, man! Animals, they're alive, not plants." I ask how they arrive at such a conclusion. "Easy, plants don't make any noise!"

A toddler tears at the wings of a butterfly. "Stop that!" his parents tell him, disturbed by the animal's suffering. Then the child grabs a branch and beats the leaves, causing debris to rain down. They say, "It's OK, let the child play, he's not hurting anyone." It is true that the leaves will regrow, not the butterfly wings, but what does the child know? The animal suffers visibly while the plant appears insensitive, but what do we really know?

One can object that toddlers behave in such a way. Let us see if cultured people, in the spirit of the moment, show more discernment. Michel Tournier, member of the Goncourt Academy and author of the *Friday* novels, noted that Robinson, shipwrecked on the island, moved through "green cathedrals" coiffed with a "fern rolled into a horn," then, encountering a "wild goat," killed it: "This was the first living creature that Robinson had encountered on the island." The learned and illustrious author would not consider plants as living beings.

Tournier, I admit, appears to be an extreme case. Even more, he is a writer, an artist, and thus does not have to be objective. If that is what he said, why is it necessary to forbid such a value judgment between plants and animals? Let us look among the scientists. Paid to be objective, cannot they, better than the general public, better than artists, forget the privileged ties we have with other animals? Would not they place the two groups of organisms on the same level, rejecting a warped perspective and recognizing the merit of plants? It would be naive to think so. On the contrary, biologists and physicians appear to me particularly attentive to preserving a central place for humans in their preoccupations and research. Humans and, secondarily, animals offer them the best opportunities for experimentation.

Contemplating Our Navel

Earth, for Ptolemy (in the second century B.C.), was the center of the universe. For scientists at the end of the 20th century and the beginning of the 21st, humans are the center of biology. We contemplate our navel; it is what we do best. In the preface to *The Meaning of Evolution*, George Gaylord Simpson (1950) at least explained why he had practically nothing to say about plants: "Specialist and non-specialist alike may note the scanty attention given to the evolution of plants, which is confessedly a serious omission in a study of the history of life. There is, however, need for reasonable brevity, and despite abundant differences in detail the principles of plant evolution are generally the same as those of animal evolution. Man is an animal, so that animal evolution is usually more interesting to him and is also more likely to have meaning for him and to elucidate his place in the cosmos. . . . It is fair also to add that I have devoted most attention to subjects in which I am most nearly competent or at least with which I have best first-hand acquaintance."

It is useful to remember that during the 18th and 19th centuries, a biological problem would not be considered adequately treated without referring to both kingdoms. Without involving ourselves in a serious historical exercise, it suffices to refer to some of the important works to verify that in Europe, plants held an honorable place in the preoccupations of biologists:

Stephen Hales, *Vegetable Staticks*, 1727

Henri Du Trochet, *Recherches Anatomiques et Physiologiques sur la Structure Intime des Animaux et des Végétaux, et sur Leur Motilité*, 1824

Henri Du Trochet, *L'Agent Immédiate du Mouvement Vital Devoilé des Sa Nature et dans Son Mode d'Action Chez les Végétaux et Chez les Animaux*, 1826

Barthélemy Charles Joseph Dumortier, *Recherches sur la Structure Comparée et la Développement des Animaux et des Végétaux*, 1832

Claude Bernard, *Leçons sur les Phénomènes de la Vie Communs aux Animaux et aux Végétaux*, 1878–79

Why did this rule, which appeared to be a natural intellectual requirement, disappear at the beginning of the 20th century, to be replaced by increasing specialization? Claude Bernard (1878–79) was partly responsible when he wrote, "The cell, identical in both kingdoms, among animals and plants, . . . demonstrates the structural unity of all living things," and "There are scarcely any important differences between the living processes of animals and plants." Finally, I find the following statement particularly suspect epistemologically: "It is of the greatest importance for general physiology to insist on the analogies between the animal and plant kingdoms." This shows an influential member of the French Academy of Sciences promoting his own discipline at the expense of objectivity. Alas, scientific specialization now has a different, powerful, and dangerous motivation: economics. Only specialized research is judged as directly applicable. The reader is warned: The book in your hands tries to revive an intellectual tradition of the 19th century.

Actually, many works in biology, whose titles give hope for a coverage of everything, are limited to a study of animals or human beings. Plants are mentioned only as an afterthought, comprising less than 8% of the text and illustrations, or are completely ignored in the following examples:

Alex Comfort, *The Biology of Senescence*, 1956

Ludwig von Bertalanffy, "Principles and theory of growth," 1960

Alain Prochiantz, *Les Strategies de l'Embryon: Embryons, Gènes, Évolution*, 1988

Stephen Jay Gould, *The Book of Life*, 1993

David M. Raup, *Extinction: Bad Genes or Bad Luck?*, 1991

Jean-Louis Revardel, *Constance et Fantaisie du Vivant: Biologie et Évolution*, 1993

Claude Combes, *Interactions Durables: Écologie et Évolution du Parasitism*, 1995 (*Parasitism: the Ecology and Evolution of Intimate Interactions*, 2001)

Dominique Doumenc and Pierre-Marie Lenicque, *La Morphogenèse: Développement et Diversité des Formes Vivantes*, 1995

Jean-Jacques Jaeger, *Les Mondes Fossiles*, 1996

Donald E. Ingber, "The architecture of life," 1998

The March 1997 issue of the French scientific magazine *La Recherche* was dedicated to the history of life. In it there is a single page concerning plants, and 25 articles on animals. In the great gallery of evolution in the National Museum of Natural History, Paris, the space dedicated to plants is ridiculously small. We understand why so many of us—children, ignoramuses, artists, mystics, etc.—are subjective, but since when do scientists have the same right?

A Bias Bordering on Injustice

The reader should not be misled. What shocks me is not that biologists choose to speak only of animals and are not interested in plants. After all, as George Gaylord Simpson said, we prefer to talk about what we know. What shocks me is that the works just mentioned pretend, at least implicitly, to cover all of biology; their titles clearly bear witness to this. They are all excellent works, but that does not solve the problem. Furthermore, their credibility lends support to the notion that true biology concerns humans and other animals.

And plants? It is agreed that they must conform to concepts developed for the animal kingdom, without a shadow of proof in support of what appears to be a simplistic zoological point of view. Many zoological centrisms are exposed in the following pages, and they impose themselves on us without our being aware of a pure and simple bias.

It was a bias bordering on injustice in the case of Barbara McClintock. In the 1940s, McClintock, at the Cold Spring Harbor Laboratory on Long Island, New York, discovered the first examples of genomic lability in the cycle of BFB (breakage-fusion-bridge) and transposable elements, a completely new genetic mechanism (Chapter 5). Why were these results, not at all idiotic, not immediately accepted by the scientific community, supposedly sensitive to such new discoveries? Without doubt it was because Barbara McClintock was a woman and of tiny stature (Fincham 1983). Another handicap was that she worked on a plant, maize (Zea mays), and the results were difficult to understand in the context of those obtained by the school of Thomas Hunt Morgan from the fruit fly (Drosophila). The latter results were genetic orthodoxy, establishing the constancy of the genome. We know from her biographer, Evelyn Fox Keller (1983), that McClintock remained misunderstood for decades. It was only after transposable elements were rediscovered in animals that her ideas were appreciated, and only in 1983 did she receive the Nobel Prize for work that went back nearly 40 years.

This is not an isolated episode. Fundamental discoveries in plant
biology are only truly accepted if they are ultimately confirmed by
animal experiments (Pierre-Henri Gouyon, personal communication). Let us be fair—some scientists see at what point our understanding of biology is harmed by the almost exclusive reliance on
animal models. The physiologist Anthony Trewavas (1982) put his
finger on the problem when he noted, "Plant scientists have yet to
face the unpalatable fact that in attempting to understand plant
development, they may have elected to examine only those features
of plant development which bear similarity to those in animals (the
production of vascular tissue, for example) but have ignored much
more interesting and novel characteristics." One of the great biologists of the 20th century, John Harper was preoccupied with the
fundamental laws of the organization of plants. He took an admirable point of departure, comparing plants to animals that are
sessile versus animals that are mobile.

For the most part, we biologists are no more free than if we were
chained to the oars of a galley. Each of us is sealed in a narrow specialty, with no possibility or even the desire to look beyond, pushed
by competition between research groups, subjected to the press of
technological advance, obliged to find grant support, surrounded
by totalitarian systems of evaluation that snuff out new ideas and
discourage a broad scientific culture. The public, which pays the
bills, is isolated from this process. The engines of this system, "pro-
ductivity" and narrow specialization, give biology the aura of a great
private grove of trees under which the humus is carefully churned by
armies of reductionist earthworms. Grand scientific syntheses, sym-
bolized by those lofty trees, are ready to take root but our blind pre-
conceptions about life create major obstacles.

The Garden and the Peasant

I wonder if, in Western culture, men and women share this biased
view of the supremacy of animals over plants. There is no need to
push the psychological investigation too far because the answer is
rather simple. Women are willingly seduced by the beauty of plants,
by the peace that emanates from them, by their perfume, by their
usefulness as food, spice, and medicine. Men clearly prefer animals,
particularly wild ones. Is this a sort of atavism? Animals evoke the

hunt, a masculine and even macho domain, whereas plants evoke the earth, nourishment, fecundity, subjects with which women are more at ease.

In my parents' time, in their social milieu, a man would rise at daybreak to go out with friends to raise the lobster traps, fish, flush out the wild boar, shoot ducks, but he would never turn his attention to breathing in the scent of a flower. Only a woman would permit herself to express such sensitivity.

Benoît Garrone, who teaches botany with me at the University of Montpellier, has shared some insights into the link between women and plants. He has noted that zoology courses are taken primarily by young men. Some of the field exercises, such as studies of birds of prey, comprise almost 100% men. In contrast, in botany courses, even in field exercises in the most rugged and wild terrain, it is women who mainly participate. The guys would say, "Botany, that's for women and gays." Perhaps they also think this of the professors.

Yildiz Aumeeruddy, an ethnobotanist specializing in traditional communities and environments in the Himalaya, was happy to share with me what attracted him to plants: "Above all they are a source of peace. My relations with plants are made simple and calm by my feelings of shedding a universal sadness. Their simple presence is profoundly comforting and calming. To observe plants is to allow them to spread their beauty over me, of forms, colors, scents. This is an unfailing means for forgetting, removing my cares and feelings of depression. To observe plants directly is not always easy. They are less easily approached than animals. Some preparation is necessary, which certain special moments of my childhood have allowed me to experience with little effort. To become interested in plants is also to join a tradition. We are spontaneously attracted by animals, but we learn to love plants." Yildiz continued, plants have "a complete vulnerability that touches me deeply, and my attachment to them is in large measure founded on a feeling of compassion, reinforced by the fact that plants are more generous than animals; they give endlessly but never ask for anything in return. I can take something from a plant without killing it, like a bouquet of flowers or a basket of fruit. No animal could ever allow such dissection."

Many women, I believe, would find their feelings about plants explained by Yildiz's words, and perhaps some men as well. Vincent van Gogh, paled by the tragedy of King Lear, said that after reading several pages, "I am always obliged to go see a sprig of flowers, a pine branch, or a spike of wheat, to calm myself" (Michel 1999). Nelson Mandela (1994), during his years of imprisonment on Robben Island, kept his sanity thanks to plants. He said, "A garden was one of the few things in prison that one could control. To plant a seed, watch it grow, to test it and then harvest it offered a simple but enduring satisfaction. The sense of being the custodian of this small patch of earth offered a small taste of freedom. . . . This was my personal means of escaping the universe of concrete that surrounded us." Not everyone shares this particular sensitivity; for others there are trees. Trees—they stand erect, they are hard, they are virile. These are possibly the reasons why forestry, even now, is a career for men. Forestry conserves some of the masculine traditions of the military: officers, generals, tidy khaki uniforms. Foresters are also much influenced by the cylinders of wood that expand over time—logs, raw materials of big business.

What do plants lack to enjoy the same consideration as animals? Are there some errant notions we can rebut? Are plants not beautiful, discreet, silent, self-supporting, and useful? Is there anything we can criticize about them? I find this proposition paradoxical: Plants are to us at the same time too familiar and yet too strange to inspire the sympathy and admiration they deserve.

The Sorcery of Omnipresence

Johann Wolfgang von Goethe (1831) wrote, "All the things surrounding us from our infancy persist forever as something common and trivial to our eyes." How well this applies to plants. They are nearly everywhere, very ordinary, too familiar to notice. How can we admire something we see every day, in the same place, year after year? How can I continue to be astonished by the plane trees lining the avenue, the brambles on a talus slope, the mosses in the cracks of the pavement, the chestnut in the plaza? Their ubiquity is a disservice to them; we appreciate plants only when they have disappeared, and that is why townspeople dote on them. See their balconies spilling over with geraniums, their apartments where philodendrons fight for space with dieffenbachias. Urban florists make their profits selling pretty, exotic weeds.

The geraniums of the city square, the begonias on the railway platform, the aspidistras in the apartment lobby—we become familiar with all these plants, seen so often. If they had been collected in some faraway solar system and brought back by an expedition from outer space, would they be more deserving of our attention? What sort of view would it take to see plants as they really are, free from the cares of our routine lives, of the dulling power of habit—perhaps a single-image random-dot stereogram? That is a two-dimensional drawing, repetitive and incomprehensible, a sort of cameo deprived of meaning, which is suddenly grasped in three dimensions by an ocular acrobatic feat that some find too difficult. Abruptly, the image becomes luminous, profound, fascinating, rich in structures unexpected and topographies unsuspected, in which our eyes, resting from the effort, revel (Ninio 1994). The privets at the post office, the wild grape in the cul-de-sac, the nettles by the spring—we see them from habit in the first degree, one could say flat, and thus they inspire nothing, not even distaste or boredom, nothing. A transformed vision could change a weed into a marvel, and we could thus find an enchantment capable of reforming our existence.

I see it as the moment when I became a botanist. At the time, as a student at the Sorbonne, I seemed more interested in zoology and paleontology. On my window ledge in the Latin Quarter, above the gray roofs, in the dusty earth of an abandoned pot, a seed germinated one morning in April and a seedling began to grow. I had absolutely no idea what it could be or how it could have arrived in such a place. I watched it grow, day after day, until it began to produce flowers, tiny but with a pure and austere beauty, then odd little fruits crammed with seeds. This pretty intruder, capable of growing and reproducing completely by itself under the oystershell skies of Paris, seemed to symbolize life where there was no beauty. Later, I learned its name, shepherd's purse *(Capsella bursa-pastoris)*. Then I left the Latin Quarter and, to nourish my newly found vision, fraternized with the most beautiful plants on our planet. Forty years have passed, but the memory of that moment of lucidity is always with me.

Beyond their omnipresence, which too often hides their beauty, plants suffer from another handicap: their mystery. That may seem contradictory, but it is like this: Plants are at the same time common and mysterious, and at least apparently unfathomable.

And the Disadvantage of Otherness

What enigmatic creatures! From the morning glories of the garden to the great trees of the Amazon, from the California redwoods to the water lilies at the edge of a pond, plants are everywhere, green, silent, strange companions with whom conversation is rare. Truly, succinctness is a rule for them. They seem motionless but only because they experience time differently than we do. We are told that they are alive, but the kids at the youth center had a point—plants are clearly devoid of the commonsense attributes of life: movement, arguments, smiles, love, crying babies. They are content merely to exist, in absolute immanence, united in their will to live, in their resolution to be themselves. "Plants," Pierre Lieutaghi (1991) said, "don't move through space like a bird or leopard. They are not mountains in motion, like an elephant. They don't come to us; they are simply there, caring, suggesting, proposing, offering patience." They cannot flee; they are vulnerable and immobile, or rather, not so inclined, to my poor eyes. They have only the motion that wind provides them, seemingly insensible, or rather, not having the sensibility that I would attribute to an animal.

Let us try to understand what trees are and we will be perplexed by their mixture of unchanging presence yet complete otherness. Recall a habit that goes back at least to Aesop's time: We are unable to keep from investing trees with human emotions and human language, to keep from seeing the human form. Their branches become hands; their crowns, heads; their roots, feet. We see in trees both friendly and menacing moods, and we believe them capable of suffering if we wound them. They are deemed to appreciate it if we talk about them, if we caress them.

From Victor Hugo to Georges Brassens, from Jean de La Fontaine to Paul Valéry, from Paul Eluard to J. R. R. Tolkien, there are many who have spoken with trees, taking advantage of their inability to answer. Why is there such a tendency to dress them up in our tawdry finery? Why must they resemble us a little? To assuage the unease we feel in the presence of these strange guardians, who exist in a time that is not the same as ours? Since the mystery surpasses us, as Jean Cocteau said, "We feign to be the originators."

In horticultural language many words come from medicine or zoology: the *heart* of a tree, its *trunk*, its *feet*, an *eye* denoting a bud, *wound, sore, healing, beheading* or topping, saying of wood that it is *veined.* The same occurs in the language of botany: *ovary, epidermis, ovule* for egg, *nerves, venation, placenta, pith, vessels, joint* of a stem, *micropyle,* etc. The opposite occurs, investing humans with vegetable traits. Those words have negative connotations, to *vegetate,* to *plant* one's feet, etc. If the physician declares that the patient in a deep coma is in a *chronic vegetative state* the verdict is more direct: The poor person is nothing more than a vegetable.

I must confess that I have also transferred human feelings to plants. At the base of a great makoré tree *(Tieghemella heckelii)* in Gabon or a meranti *(Shorea)* in Sumatra, I cannot help but sense a sly haughtiness, as if the tree is amused by the strange new organism, bilateral and noisy, who moves about at the base of its trunk. He looks up, trying to identify the leaves, grovels for fruits in the litter, then suddenly lowers his pants to relieve himself, showing that he lives in a very different time scale, one in which the problems of waste removal have not been solved very elegantly.

Do not think that it is sufficient merely to study plants in order to really understand them, or even to see what they are. You can be a butcher, bonesetter, florist, botanist, pineapple or olive cultivator, creator of bonsai, manager of a vegetarian restaurant, or simply an amateur with bouquets or fruit salads without capturing one iota of what Marshall Darley (1990) has called so humorously "the essence of *plantness.*" Plants are so constructed that there is no need to understand them in order to use or admire them, or certainly, to destroy them. You can easily fell a tree and cut it into logs—you neither need to know what it is nor what it thinks of you!

Is dialogue with plants thus impossible? Are plants so different from us that we must abandon any hope of penetrating their privacy and making progress in understanding their true nature? Certainly not. One of the objectives of this book is to reveal things through an inner understanding in an attempt to comprehend plants. The first step in this direction is to recognize that in certain ways plants disclose a great wealth of sympathy. For example, they are of great importance in providing beauty, which has been recognized since ancient times. Jesus spoke, "Consider the lilies of the field, how they grow; they toil not, neither do they spin. And yet I say unto you, that even Solomon in all his glory was not arrayed like one of these" (Matthew 6.28–29). What a pleasure to imagine Jesus with his nose in a lily, admiring it, and how worthy, taking a botanical moment in a life consecrated to all else.

I share an obvious truth in recalling that artists, writers and poets, painters and printmakers, have always found inspiration in contemplating plants:

> Harvest moon
> On the bamboo mat
> Pine-tree shadows
>> ENOMOTO KIKAKU (1660–1707), a disciple of Bashō,
>> as quoted by Bownas (1964)

> In timeless forest
> A great tree is felled
> A vertical void
> Trembles in the form of the bole
> Near to the fallen trunk

> Look, look, birds
> The place of your nests
> In the high memory
> As long as it still murmurs.
>> JULES SUPERVIELLE, *Gravitations*, 1937

In our time, trees have become symbols of our respect for nature, and ecological movements are described as green. This is certainly better than describing a political party as red as a fish, blue as a kingfisher, or spotted like a giraffe. Rendered beneath the branches of an oak, justice can pretend to be impartial; under the belly of a camel or between the arms of a gorilla, it would be . . . more than meets the eye. We see that plants still play an important role; we also see very clearly that plants and animals have very different attractions for us.

A Bit of Psychology

The interest we have in animals, from infancy on, comes from a feeling of identification; they fascinate us to the degree that they resemble us. Furthermore, we establish among them a sort of hierarchy, not formal but very real. A primate provokes more interest than a rat, a mammal more than a fish, which in turn attracts more attention than an earthworm. We do not venture outward too far; for

PLANTS, ANIMALS, AND HUMANS

most of us a nematode or ctenophore evokes nothing more than a Captain Haddock oath.

Interest and curiosity are governed, at least in part, by an emotion that culminates in humanity, recognizing itself as a species, and that diminishes if the living being is different. Thus your stock falls if you have paws instead of hands, scales in place of skin, if you live in water and are not even capable of crying if made to suffer. At the very least, there must be mobility to allow identifying, even with an insect. The animal that we envisage always has the aspect of a human, giant or homunculus, according to the situation. Aesthetics can play a role in this attractiveness, but only secondarily.

Animals are more familiar to each other than to plants, and connivance between us animals allows us to understand each other. If I place some seeds on my balcony during the winter, I know that the blue tits will come, and they know that I will leave something outside because this is an annual rite. They know that I do not bother them and will not open the window while they are foraging. I know that they await springtime, as I do myself in order to sow the morning glories when the birds return to hunt for caterpillars in the big white oak. No doubt, I could be deceived by them. Nevertheless, I have a conviction that errors of interpretation committed in regard to a blue tit—compared to a bunch of daisies, for example—are counterbalanced by a capacity for prediction since I am a part of the community of animals.

To the reader who is astonished that a botanist speaks with animals, I say I am not a zoologist or a hunter, I agree, but better than that, I am an animal. On the other hand, plants represent absolute otherness to us, and what attracts some of us toward them is at its heart a complex mixture of feelings. That attraction has a role, marginal but real.

Carnivorous plants attract the public. At botanical gardens they draw crowds, which is notable because they are rather dull and stern in appearance *(Drosera, Sarracenia, Dionaea, Drosophyllum)*. They have a character that makes them attractive, however, and it is an animal character.

One does not need to be a psychologist to discover the reasons for our fascination with plants that visibly grow *(Phyllostachys)*, plants that move *(Mimosa)*, plants that dance in response to noise *(Codariocalyx)*, plants that stink of excrement *(Aristolochia)* or of a cadaver

(Stapelia), plants that speak *(Hernandia)*, etc. Toward plants, humans "accord only a vague and vain curiosity, only if, condescendingly, one detects something human in their attitude" (Brosse 1958).

What is essential to their attractiveness is moreover not their human or animal characteristics but their extreme visual and olfactory beauty; Jesus, we have seen, was sensitive to this. Plants also owe their power of attraction to the profound feeling of peace that they inspire. To watch an animal creates tension because we know the moment is fleeting. To observe a plant engenders serenity; it is time itself that becomes visible. The plant's growth is slow but perceptible with close attention. We renew our connection with the peaceful rhythm of time that prevailed during our infancy.

I ask the indulgence of those who hold different feelings. Please consider my point of view. I write as a botanist, an observer of the natural world, not as the supervisor of a stud farm or a farmer raising weeds!

In all the vegetable world, trees have yet another attractiveness, much more mysterious and grand. Because of their long lives, because their image is shaped by eternal forces, because their "ascending verticality" (Vieljeux 1982) unites earth with sky, crossing the domain of humanity, trees "seem the support most appropriate for all cosmic dreaming" (Brosse 1989). For Mircea Eliade (1949) the tree is the symbol of the cosmos, a role it has always had and still has in most religions. Eliade wrote that confronted with a tree, humans "are capable of reaching the highest spirituality, in comprehending this symbol they succeed in living the universal." It does not seem that animals or we ourselves have ever attained such dignity.

Let us not dream: The tree is perhaps the symbol of the cosmos, but no matter what, protected only by its trunk, it is capable of being destroyed in half an hour. Such vulnerability is disturbing to those who love trees; plants need our protection. I aspire to show that plants are not inferior to animals, as often believed, and are essential to the degree that we humans cannot exist without them and would not survive their disappearance.

Comparing Plants and Animals

To transcend their otherness, to try to understand what plants are, what kind of being a plant is, I compare them to animals, with which we are more familiar. This comparison has been attempted before, but rarely, in the scientific literature. Since the 1970s, comparison of plants and animals has rarely risen above the level of a brief mention. The only works significantly devoted to the subject are, to my knowledge, those of van Steenis (1976), Valentine (1978), Southwood (1985), Walbot (1985), Darley (1990), Prost (1992), Reeves and Obrenovitch (1992), and Thom (1993). These works are brief and, in a way, indirect. The authors sense the importance of the question but refuse to dedicate themselves to it fully. The results are incomplete, even superficial. The best is that of Darley, whose article of three pages is exceptionally dense.

Apart from these several titles, the need for discussion of the two kingdoms nevertheless preoccupies many biologists. I have mentioned Harper and Trewavas, and I finish this introduction with a quote from one of the greatest plant physiologists of the 20th century, Kenneth Thimann: "The similarities and differences between the development of plants and animals, although very clear in appearance, remain to be defined with exactitude in physiological terms. It is possible that the cross-fertilization between the science of plants and that of animals constitutes for the future the most fruitful area of research."

Before I launch into the heart of this subject, I would like to state my conviction that the task is difficult, that plants do not readily divulge their secrets. Understand that this is a task requiring much time and effort, and multiple sensitivities. Reason does not replace intuition, but we scientists have not gone down that latter road very far or very quickly. Others have preceded us—gardeners, poets, mere hikers, monks, dreamers, scholars, bonesetters, mushroom collectors. They have their say in the following pages; the task is difficult but we have allies who understand at what point they surpass us in insight. Our collective perception of plants is at a crude level, and the sensitivity of such persons can prove to be more penetrating than the reasoning of a specialist.

What are plants? Despite the important place they occupy in our landscapes, they are often ignored or scorned by modern biologists, by loggers and the old foresters, by people in general. Does that place us on a contrasting quest to understand plants? Plants please most women, they pleased Christ and the Buddha, the last emperor of China, the scholars of earlier centuries, and they have pleased innumerable artists and philosophers, including Goethe, Cioran, Colette, Valéry, Mandela, Dürer, Giono, Hugo, and Rilke. First, what do plants resemble and how can we comprehend the forms they have adopted?

CHAPTER 2

A Visit to the Landscape of Form

Nothingness has an architecture that makes real demands on things. Every form, every pattern, every existing thing pays a price for its existence by conforming to the structural dictates of space.

PETER S. STEVENS, *Patterns in Nature*, 1974

Look, the trees *are*.

RAINER MARIA RILKE, *The Duino Elegies*, 1923

. . . a plant is a song whose rhythm deploys a definite form and within space displays a mystery of time.

PAUL VALÉRY, *Dialogue of the Tree*, 1943

I T IS NOT EASY to consider form without attending to habits of speech and the constraints of language, and their bearing on scientific communication. The science of form, or morphology, seems terribly obsolete or not even a true science to many contemporary scientists. Following Thom (1972), Boutot (1993) has shown very well the reasons for the distancing of contemporary science from morphology. The principal reason is that form is fundamentally qualitative, unmeasurable in length, mass, speed, or temperature. Not being measurable, form is not a precise object for scientific investigation. That is the view of those who think that science is only quantitative. I am reminded of the pronouncement attributed to Ernest Rutherford, "Qualitative is nothing but poor quantitative," which conveys the contemptuous attitude of quantitative science.

The theories of morphology "are written on the margins of the technical and scientific universe, of which they shatter most of the idols," celebrating in a manner "the lost facts of science and philosophy." These theories generally place contemplation higher than

41

action, "enormous audacity in a century when research is officially registered, subjugated to the demands of output or administrative routine" (Boutot 1993).

Whence Form?

I would like to explain my belief that knowledge of the form—of any object, whether plant or animal—gives access to more essential information than quantitative investigation can provide. Faced with a plant, I learn more from observing its form than from determining its alkaloids, working up its mineral constituents, sequencing its DNA, measuring the different microclimatic variables surrounding it, elucidating the homeotic genes that control its flower structure, and so on. Certainly, the sum of the information provided by all these quantitative approaches can be synthesized, if someone can be found to do it, to establish a body of knowledge of unmatched value concerning the plant in question. But such an approach is utopian; it is much more efficient to grasp form than to be limited to analysis. This is especially so if the synthesis is not accompanied (as so often the case) by simple observation of the plant.

Does there exist in the plant a quantity of profound importance not inscribed in its form? I do not think so, and nothing in scientific texts, ancient or modern, comes to my mind that demonstrates the contrary. Form is an integrator of internal tendencies, itself offering a powerful synthesis. I believe that the science of form is not obsolete; it has its place in a "philosophy of nature in contemporary life" (Boutot 1993). Together with Thom (1972) I admit, "The pageant of the universe is an unceasing movement of birth, of development and destruction of forms, and that the object of all sciences is to predict this evolution of forms and, if possible, to explain it."

I can imagine what Rutherford's opinion would be if the quantitative that he liked as verification added up only to an infantile or embryonic qualitative. Then, all the sciences could be recapitulated simply, and most qualitatively, as form. This justifies a comparative study of animal and plant form. Let it suffice to say that in biology, form is the product of growth, which simultaneously involves time and space. The reader who is interested in birth, growth, and death will find them here associated with form, like truffles beneath an oak. So that discussion may occur in the proper context, it is necessary to recall that plants and animals rely on different sources of energy.

Capturing Energy

A living being, whatever it is, needs energy—a permanent, daily need that is certainly common to both animals and plants. This requirement for energy suggests an engine that needs fuel, but the metaphor is false. The engine stops without fuel, running again when it is supplied. Without energy, an organism is condemned to death after only a brief delay, as if a car engine destroyed itself several hours after running out of fuel. "The organism cannot be dissociated from the energy that feeds it" (Passet 1979). An organism needs energy not simply for functioning but for existing. Whether plant or animal, capturing energy is imperative.

However, the similarities stop there. Plants and animals gather energy by different means, and the forms of energy they use are different. This divergence in mode of energy capture leads to distinctive and important traits that separate plants and animals. It is a fundamental difference; others are mainly consequences of this basic choice between two sources of energy.

Plants, Vast Fixed Surfaces

Everyone knows that plants use energy that comes directly from the sun. This is energy carried as photons, a radiant energy of high quality (Dyson 1971). Its intensity is rather weak, however, only about 1 kilowatt per square meter on average (Rudaux and Vaucouleurs 1948). Plants use this energy rather inefficiently. A consequence of the relatively low intensity of this energy source is that plants must augment their linear and surface dimensions to the detriment of their volumes. Another consequence is that solar absorption must continue as long as possible, stopping only at night.

Because solar energy arrives at the absorbing organ from all directions, the surface's proper placement guarantees better capture. In other words, the surface placement should not be disadvantageous to the plant. Moving large surfaces is cumbersome, and the plant's fixed position on the ground provides the additional advantage of allowing a supply of water from the soil. Since the intensity is low, producing a surface adequate for capturing solar energy is of paramount importance.

Thus, in spite of its modest volume, a plant must produce huge subterranean and aerial surfaces supported by a very large linear infrastructure. Even the massive trunk of a large tree represents only a small living skin covering the dead wood; the living mass is very small in relation to its surface. Measuring this plant surface is not easy. For a tree, we must count the numbers of branches and leaves, measure the areas of leaves and branch surfaces, then add those to that of the trunk. Such work has only been completed for a few young, small trees:

> 340 m² for a young chestnut *(Castanèa)* 8 m high
> 400 m² for a small oil palm *(Elaeis)* 3 m high
> 530 m² for a spruce *(Picea)* 12 m high

We lack the allometric rules that would allow us to use the measurements of a young tree to estimate the areas of a large adult. What would be the surface area of a tree 40 m high? An estimate of 10,000 m² (1 hectare) is certainly not an exaggeration, probably an underestimate. Besides, this external surface is only one aspect of the problem. It is also important to consider the internal surface, which allows gas exchange in the substomatal chambers inside the leaves, a surface 30 times greater than that externally. Thus for a young orange tree *(Citrus)* with 2000 leaves, the external surface area is 200 m² and the internal surface raises the total surface area to 6000 m² (Vogel 1988).

Research is more difficult and results sparser regarding root surfaces. The root surface of a single rye plant *(Secale cereale)* is as much as 639 m²; the underground surface is 130 times greater than that above ground. The roots, placed end to end, would be 622 km long (Dittmer 1937) with a daily increase of 5 km (King 1997). Adding the absorbent root hairs, the numbers become enormous: 10,620 km long with a daily increase of 90 km (Figure 2; King 1997).

The two ratios, 30 and 130, may not hold for all plants. If we apply them to a large tree with 1 hectare of external surface, however, the internal leaf surface would be 30 hectares; the root surface, 130 hectares; and the total surface of exchange with the environment, 160 hectares! Only part of that surface is dedicated to energy capture. The essential point is that plants are immense surfaces.

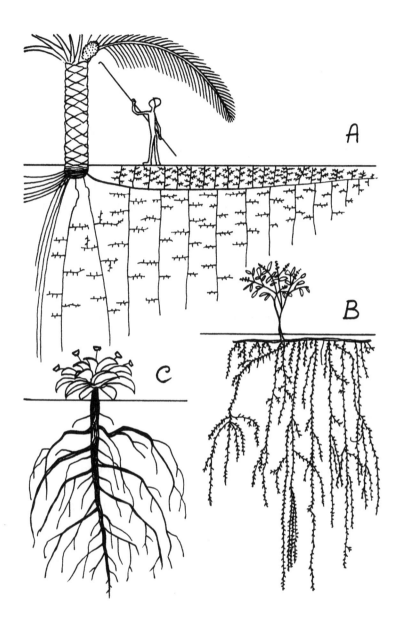

Figure 2. The surface areas of roots exceed those of the aerial portions of plants. (A) An oil palm, *Elaeis guineensis,* one of whose 4000–8000 horizontal roots is shown at the right of the trunk (Jourdan 1995). (B) A snowberry, *Symphoricarpos orientalis* (Caprifoliaceae), the aerial portion of which is less than 50 cm high, yet the root system plunges to 150 cm (Pelton 1953). (C) Balsam root, *Balsamorhiza sagittata* (Asteraceae), the aerial portion of which is about 5 cm high whereas the roots are more than 2 m deep (Weaver 1919).

When necessary, plants also know how to make structures with volume: an orange, an onion, a macadamia nut, a coconut, a hazelnut, a yam, a rutabaga, a pumpkin, an avocado. These show that plants know how to put special structures or organs under a protective surface of reduced area. Ribald zoological metaphors often make fun of such organs: the eggplant, the obvious sexual anatomical associations of the double coconut *(Lodoicea maldivica)*, nuts as slang for testicles, acorns (*gland* is French for acorn and the word has a sexual connotation), etc. Certainly, the evocations of animal form in spherical structures of plants are for mobility and dispersal. Not too stupid, these plants.

Animals, Small Mobile Volumes

Animals capture energy from food or prey, first by their mouths and then through their digestive tracts. In terms of quality, this food or chemical energy is less than that of the sunlight captured by plants but is compensated for by the amounts obtainable. Compared to plants, animals do not need to feed themselves all day because their food or prey contains so much energy. In contrast to plants, they capture energy efficiently.

[Even though I have studied the functional ecology of tropical plants for many years, particularly adaptations of understory plants for energy capture, I am struck by this insight. Perhaps it is because we use separate energy currencies for different purposes and do not often apply the conversions for direct comparison. So we use Calories (kilocalories) for food, and photon flux or energy flux (watts, or joules per second) for sunlight. The following example makes the author's point very clear. Suppose you have a quick, 10-minute lunch —cheeseburger, French fries, chocolate shake—about 1500 Calories. Now, imagine that you are a plant with a large beach umbrella (about 2 m^2) for capturing sunlight. At midday in Miami, radiation strikes the surface at about 1 kilowatt per square meter, and half that can be absorbed for photosynthesis. One kilojoule is equivalent to 0.239 Calories. If the umbrella were a perfect energy absorber, it would take about 105 minutes to capture the energy equivalent to the 10-minute lunch. However, the real difference is much greater than that. The lunch can be processed at about 30% efficiency. If the energy-capturing umbrella were a plant canopy, the efficiency would be much lower: There are reductions when the sun is nearer the horizon, light is reflected by leaves (and transmitted by leaves and the canopy), and most energy is lost during the biochemical process of photosynthesis. The plant's efficiency is thus less than 3% measured against the light shining on it, and 10% that of the digestion of the lunch. It would be necessary to hold the umbrella up to the sun about 2 days for the equivalent assimilation of energy. It is better to have a much bigger umbrella, keep it pointed at the sun, and stop walking around! —translator's comment]

Unhappily, neither food nor prey generally offers itself for entry into the digestive tract. An animal must procure it, which requires mobility. That necessitates a modest surface area because the ability to conceal oneself is inversely proportional to surface area. Minimizing surface and size favors volume. That puts the exterior of the body a short distance from the source of energy and in line with the digestive tract. The form is very much that of a sphere. Spherical form provides a maximum of volume within a minimum of surface. Add to that the double requirement of finding prey and avoiding becoming prey to others; these are powerful constraints, promoting the evolution of animals that optimize their volumes. Their mobility increases with size, up to a limit.

Animals are thus essentially volumes covered by small external surfaces. For mammals, ranging in size from a mouse to a bull, the relationship between surface and volume can be envisaged as a cube. This is the reason that some zoologists have used the cube as a model for animal form (Figure 3) to simplify calculations relating surface to volume. Certainly, many animals are exceptions; many have body shapes more elongated than a sphere. Some profit from an elongated form by hiding among plants—pycnogonids, *Phyllopteryx*, stick and leaf insects (Figure 74, Chapter 5)—and others—millipedes, worms, serpents—are in general rather slow.

With Vast Internal Surfaces

From necessity, animals also know how to create vast internal surfaces. Our lungs only hold an air volume of about 6 liters but their 30 million alveoli expose a total surface of 100 m^2, that of a nice apartment with a balcony. A better example is provided by the digestive tract, through which food energy is assimilated. As Hladik (1967)

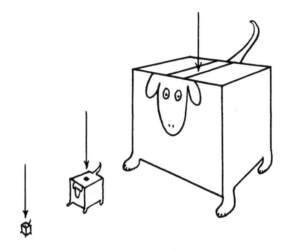

Figure 3. Cubic models of mammals. The dimensions of the cubes are, left to right, 1 cm, 10 cm, and 100 cm. Corresponding volumes are 0.001 liter, 1 liter, and 1000 liters. To permit the cubes to fill at the same rate, openings have been cut into the tops; their areas are 0.001 cm^2, 1 cm^2, and 1000 cm^2 (Chivers and Hladik 1980).

has shown, the digestive tract of a monkey, typical of mammals, is lined with villi (cellular projections) visible to the naked eye (Figure 4). These are covered with microvilli 1–3 μm long. In turn, these microvilli are covered with tufts of branched filaments, which are covered with a convoluted layering, the glycocalyx. The projections of the latter are a few nanometers (billionths of a meter) long. These four levels of surface projection—villi, microvilli, filaments, glycocalyx—together form an enormous surface for contact with food particles. As for plants, it may be premature to accurately guess this surface, estimated by Hladik as "immense." Instead of thinking of the surface as two- or three-dimensional, it may be better to describe it as fractal (Jean-Marc Lévy-Leblond, personal communication).

The microvilli, covered by tufts of filaments, have a curious plant-like appearance, the best possible strategy for increasing surface area. Is this appearance pushing the plant analogy too far? Hladik noted that the glycocalyx, situated outside the cell membrane but constructed of polysaccharides from the cytoplasm, is comparable to the cellulosic wall of plant cells, but this can be no more than a coincidence. On the contrary, a functional homology unites the digestive

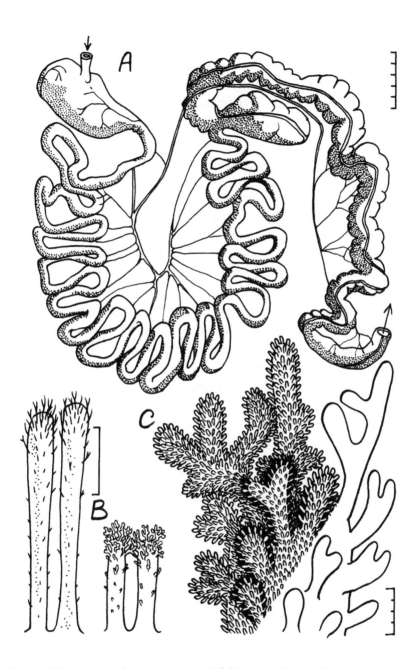

Figure 4. Digestive surfaces of an animal. (A) Intestinal tract of a monkey, *Cercopithecus cephus;* scale, 5 cm. (B) Microvilli covering the internal surface of the small intestine, each villus with a tuft of branched filaments at its tip; scale, 1 μm. (C) Tip of a microvillus, with its branched filaments revealed on the left, and the glycocalyx, formed by small bumps side by side, on the right; scale, 400 angstroms or 40 nm (Hladik 1967, Chivers and Hladik 1980).

surface of animals with the external surface of assimilation in plants. The two surfaces are equivalent in terms of energy capture (Figure 5). Animals are confused plants, turned inside out like a glove, with infolded leaves and roots in their digestive tracts. Plants are fantastic animals, their insides turned out, bearing their entrails like feathers.

Consequences of Growth

Since Euclid, we have known that growth of any object augments its volume in relation to its surface area. Starting with a linear segment and doubling its length (Figure 6), then expanding it in two and then three dimensions, demonstrates this relationship. Much later, Galileo demonstrated the physical consequences of such growth in terms of the strength of material. We have seen that different modes of capturing energy have led plants and animals to adopt different forms. We will now see how these constraints have been reinforced by the influence of growth in defining these forms, in linear, surface, and volume dimensions.

Fundamentally a volume, an animal easily accommodates the effects of growth; little change in form is required. In terms of geometry, we can say that animals remain homotopic during their development. In reality, this is only approximate; a baby has a much larger head than an adult relative to its body.

Essentially a surface, a plant would become too voluminous by remaining homotopic. Compromised in such a way, plants would not have enough photosynthetic surface and could not survive. Thus plant growth requires dramatic changes in form. Plants win this perpetual war with homotopy using two complementary strategies.

The first strategy is to take a form that elongates as quickly as possible compared to a sphere. This is obtained by dividing growth between specialized axes that fill space: trunk and branches above the ground, taproot and lateral roots beneath (Figure 5). For a plant, apical meristems are the result and have their origin in the need to avoid volumetric growth yet produce trunks, branches, and roots. These structures bear the surfaces—lateral roots and leaves—the former subterranean homologues of the latter. To the extent that a growing plant requires branching, both above and below ground (Figure 5), to fill three-dimensional space without the limitations of volumetric growth, it fills the space with a surface complexly folded

Figure 5. Homology between the external (assimilating) surface of a plant and the internal (digestive) surface of an animal. Above, two developmental stages in a plant. Below, two stages in the development of an animal, in which the digestive tract is illustrated in a simplified form. The animal inhabits only the aerial milieu, whereas the plant inhabits two environments, aerial and subterranean.

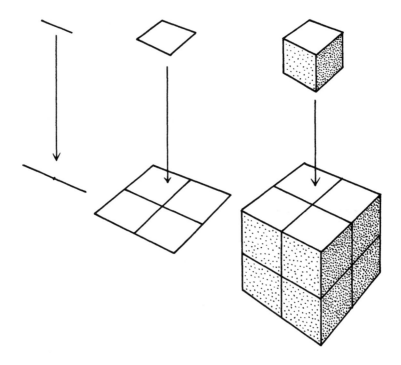

Figure 6. The differential effects of growth. Multiply the dimensions of a linear segment (left), surface (center), and volume (right) by 2. The segment is doubled in length, the surface is multiplied by 4, and the volume increases by a factor of 8. Biological growth is directly concerned with such differential effects since it occurs in linear segments (the twig of a plant, the tail of a snake), surfaces (a leaf, a wing of a bat), and volumes (an apple, a skull).

in on itself. Thus its volume decreases in relation to linear and surface growth. In this way plants resemble fractals. How do plants and bureaucracies resemble each other? In one respect, both tend to branch out as they grow.

A second strategy to avoid homotopy requires the function of another kind of meristem, the cambium, to change the material of growth. As stems and branches grow in diameter, the original tissue of large, fragile parenchyma cells is replaced by rigid, denser, less vulnerable material: wood (Figure 7). This change of tissue prevents branches and old roots from becoming too voluminous. Additionally, it satisfies the mechanical constraints resulting from Galilean scaling. Wood, mostly near the base of a tree, adds to stability and

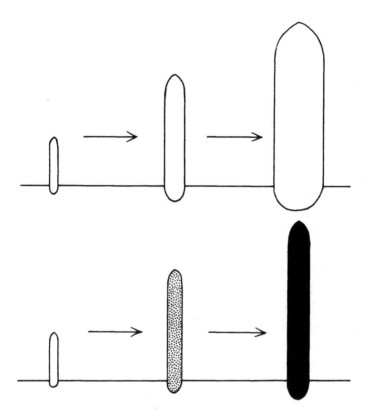

Figure 7. Changing materials of growth. Above, the original tissue is a fragile and bulky parenchyma. Below, its progressive replacement by stronger material (wood, for example) allows a relatively smaller trunk diameter.

allows great trees to remain erect. We can see with certainty the stage at which a plant becomes an initiate of Galileo.

Capturing energy certainly has a role in determining the forms of all living creatures. For those who live directly on solar energy or those who obtain their energy from food, the forms will necessarily be very different. Some will be mobile, and others fixed in place.

The Structure of Space

In all cases, the strategy of energy capture is not the only factor that determines form. Living forms are also controlled by the surrounding space, by relationships they have with the structure of that space. Since Euclid, everyone knows that space has three dimensions. It is not a question of a way of speaking or an arbitrary convention; mathematicians agree that space is objectively three-dimensional and that the three dimensions are interchangeable (Figure 8A). Time is the fourth dimension but it exists outside space. This is true of Space with a capital S, but on Earth and in its environs, in the familiar space important to organisms, the three dimensions cease to be interchangeable. In the history of life, particularly in the splitting of the plant and animal lineages, this loss of equivalence in spatial dimensions is an essential fact. Once verticality is influenced by gravity, that dimension ceases to be the same as the others. It is easy to see the difference. An infinite number of horizontal lines can pass through a point in space, but only one vertical line can (Figure 8B). Any two vertical planes can pass through the point, but horizontal planes are always parallel (Figure 8C). Vertical and horizontal extensions are different. If east and west are equivalent, like north and south, the zenith and nadir, on the contrary, differ because of the force of gravity from the Earth's great mass. We say that the horizontal plane is isotropic whereas the vertical plane is anisotropic.

Without involving physics (not my area of competence) I can only say that it is legitimate to consider that our space, as empty as it is, possesses structure. Without speculating about different spaces with unknown structures somewhere else in the universe, let us try to understand how our three-dimensional space constrains inert objects as well as living creatures, plants or animals, biologists or mathematicians. This is space horizontally isotropic locally and vertically anisotropic, at least at the scales profoundly affected by gravity.

The Scale of Natural Phenomena

At the surface of the Earth, and therefore of a globe, vertical lines converge and horizontal planes cut across each other, contradicting what I have just written. However, this observation is not important for organisms. Did Euclid know that the Earth is round? Euclidean space describes conditions for plants and animals better than a sphere (Figure 9). This leads us to divide the dimensions of life into two domains separated by a soft boundary defined here as a hiatus:

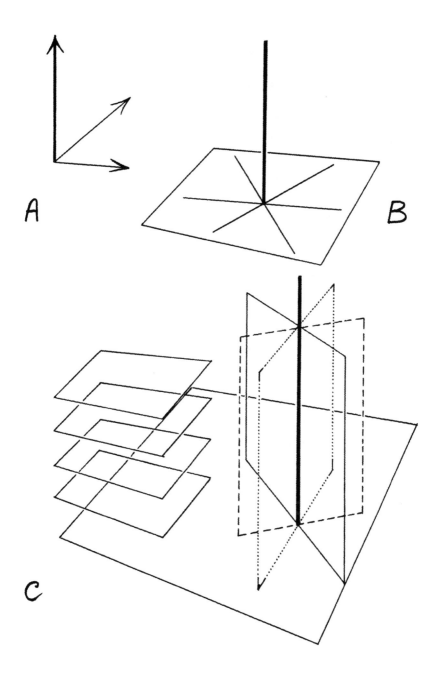

Figure 8. The three dimensions of space. (A) Vertical is not the same as the other two because it is influenced by gravity. (B) An infinite number of horizontal lines can pass through a point, but only one vertical. (C) Vertical planes can bisect the point whereas horizontal planes are always parallel to each other.

microscopic	0–100 µm
hiatus	100 µm to 1 mm
macroscopic	1 mm to 1 km

Is there a relationship between the form of an object and the structure of space? If such a relationship exists, does it depend on the object's size? I propose a crude but simple method: Describe the form of an object in a global manner, by its polarities and symmetries in the three spatial dimensions (Figure 10A). Under these conditions, an object that shows no regularity is called amorphous. An amorphous object, without symmetry or polarity, is completely independent of spatial structure. Such objects are typically nonliving and inert, such as an asteroid or a stone broken from a cliff (Figure 10B). Most often, amorphous objects result from nonreproducible processes, distinguishing them from forms with triple polarity (Figure 10C–F).

What about nonamorphous objects of macroscopic dimensions (1 mm to 1 km)? Before going much further it is useful to notice that the plants, animals, and nonliving objects with which we are most familiar are in this size range. Thus we can discuss them together; we need not distinguish the living from the nonliving prematurely. D'Arcy Thompson (1917) remarked that in "innumerable examples forms are not specifically linked to the characteristics of living organisms, but are manifestations of varying complexity of a simple physical law. . . . The living, the inanimate, us, the occupants of this world, and the world where we reside, and all sorts of unknown things, are all explained by physical and mathematical laws." With the support of this great scholar, I now move into the relationship between form and space, convinced by his authority and eloquence.

Relationship Between Form and Space

In space, a large and nonamorphous object generally has vertical polarity as a result of the force of gravity, for example, a cat or a bottle (Figure 10C, D). Exceptions occur, such as objects made to rest at any point on their surface (Figure 10E, F); the vertical is typically marked by symmetry. What about the horizontal dimensions? Why do they often have symmetry and polarity, like the cat, or an infinity of symmetries, like the bottle? Clearly, some distinction should

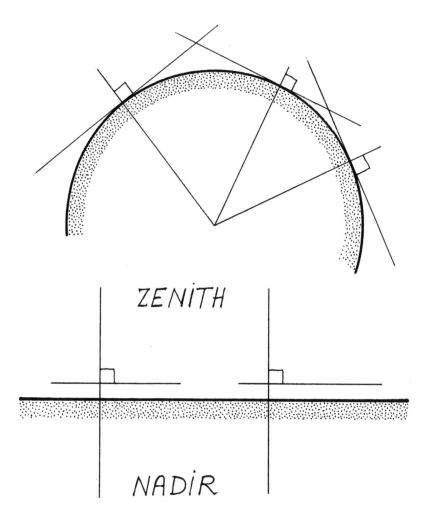

ZENITH

NADIR

Figure 9. The scale of phenomena. Above, vertical lines converge in a sphere and horizontal planes cut through each other. Below, but at the scale of interest to plants and animals, the terrestrial surface can be considered a horizontal plane.

be made between these two sorts of objects, but what? Our experience with cats and bottles naturally leads us to distinguish between mobile and immobile objects. I propose to trace a different boundary between mobile and immobile objects, which have links to the vertical on the one hand or to the horizontal on the other. What are these links?

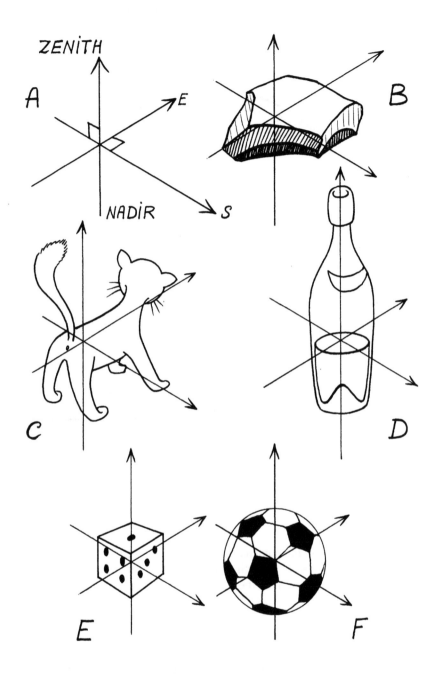

Figure 10. Symmetries and polarities. (A) The three spatial dimensions (E, east; S, south). (B) An amorphous object, a stone or asteroid. (C, D) Two nonamorphous objects marked by vertical polarity resulting from gravity. (E, F) Two objects indifferent to whatever point on their surface they rest and marked by vertical symmetry.

Objects that move vertically (a hot-air balloon or a rocket), grow vertically (a palm), or that have some vertical function (a well or candle flame) are symmetrical in their horizontal dimensions (Figure 11). In other words, they are radially symmetrical in the vertical axis, projecting almost the same image horizontally no matter what the position of the viewer.

Other, quite different objects (Figure 12) grow horizontally, such as a road or the leaf of a flamboyant tree *(Delonix regia)*, or move horizontally, such as an automobile or terrestrial animal, or have some function linked to the horizontal, such as a pair of spectacles or a rake. All these objects have some form of symmetry in one of the horizontal dimensions and a polarity in the other. Remember that in the isotropic horizontal plane, the position of the object indicates two directions. Thus each of the objects in Figure 12 has an anterior–posterior polarity in one of the dimensions, defined by its direction of development, function, or movement. In the other dimension, at right angles to the first, bilateral symmetry allows it to move, function, or grow in any direction. This is only relevant in our familiar space. Since it is not set in a horizontal plane, the spaceship *Mir* does not have dorsiventrality or anteroposterior polarity, nor is it bilateral. With its aligned cylindrical modules and its solar panels facing the sun, there is something curiously plant-like about it (Bruno Corbara, personal communication).

The situation in which each of the three dimensions of a form has polarity is rare and, frankly, marginal. We find examples in some unusual vehicles that have particular constraints (Figure 13), such as the Venetian gondola, or that are incapable of functioning correctly, such as the Blohm und Voss 141, an asymmetrical airplane. These are objects constructed by humans, and no natural, inert object has triple polarity—such objects are amorphous.

Triple polarity in organisms is rare. The only group of numerical importance is the gastropods—snails, periwinkles, *Murex*, whelks, *Limulus*—in which the spiral shell breaks the bilateral symmetry (Figure 14). The spiral is a special case of symmetry (Éric Térouanne, personal communication), however, and loss of strict bilaterality is reconciled by the slow movement of these mollusks. What advantage does bilaterality confer on organisms? Cephalopods, such as the squid, octopus, chambered nautilus, paper nautilus, and cuttlefish, are the quickest moving of the mollusks. They all have the symmetry

Figure 11. Vertical objects with vertical growth, vertical movement, or function marked by verticality. They have vertical polarity because of gravity, with radial symmetry along their vertical axis such that whatever the angle from which they are observed horizontally, the image is the same, or nearly so.

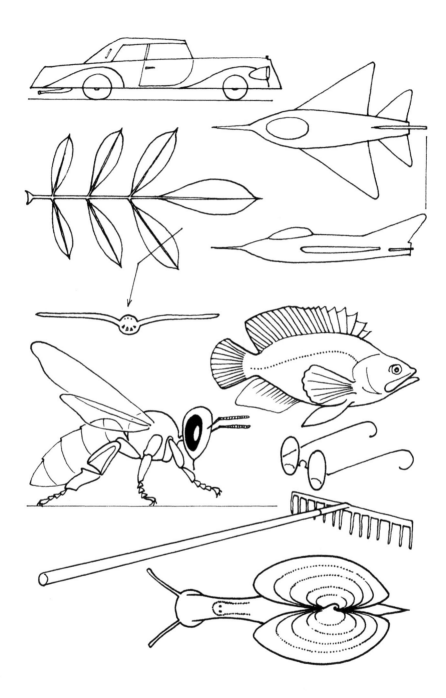

Figure 12. Horizontal objects with horizontal growth, horizontal movement, or a function marked by horizontality. They have vertical polarity because of gravity, with anteroposterior polarity defined by the direction of growth, movement, or function, resulting in bilateral symmetry.

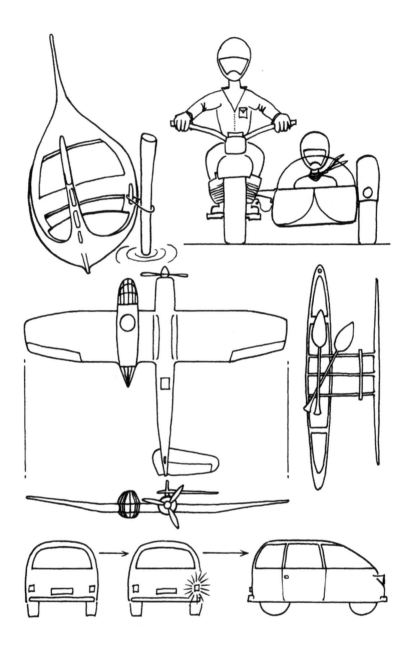

Figure 13. Mobile, asymmetrical objects: a Venetian gondola, a side-car, a single-engine fighter (Blohm und Voss 141) that flew briefly for the Germans during the Second World War (drawn courtesy of the German Museum, Munich, and the Aeronautical Museum, Bourget; Nowarra 1993), and a Polynesian pirogue. Below, in a symmetrical automobile, loss of bilateral symmetry announces an imminent change in direction.

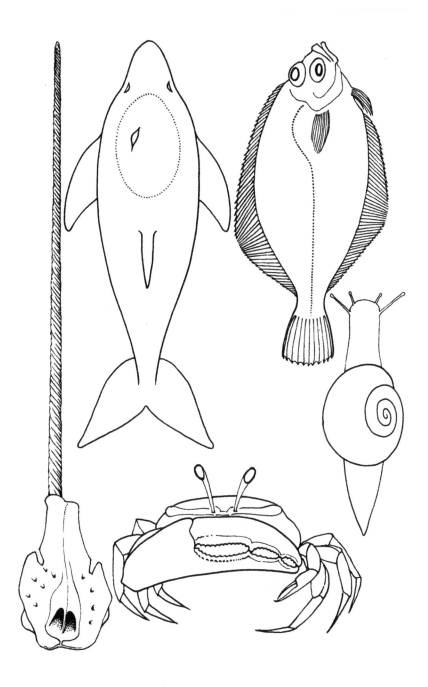

Figure 14. Examples of animals in which bilateral symmetry is only approximated: tooth of a narwhale (Grassé 1955), vent hole of a sperm whale, eyes of a flatfish (Policansky 1982), shell of a snail, and claws of a crab. These are all small alterations in symmetry in response to particular constraints.

and polarity normally associated with fast movement, as does the opistobranch *Berthelinia* (Figure 12). Of the three polarities in gastropods, the right–left polarity is more superficial than the other two.

Other examples of triple polarity, or rather the approximation of bilateral symmetry, are found among animals: the dissymmetric jaws of some collembola, the wing coverings of crickets, the enlargement of one of the claws of many crustaceans—lobster, crab—and violinists!, the copulatory arms of some octopuses, the sexual organs of the fishes *Anableps* and *Gulaphallus* (Franc 1968). In these fishes the male seizes the female with its priapus in a sort of embrace, first on the right, then the left (Parenti 1986). Add to these examples the skull of a pleuronectid fish such as a sole or halibut, modified by the migration of an eye to the other side (Policansky 1982). Other examples include the asymmetrical coloration of amphibians such as salamanders or poison arrow frogs, or of the African hunting dog *(Lycaon pictus)*, ears situated at different heights in the barn owl, the strongly curved beaks of crossbills *(Loxia)*, the vent hole of a sperm whale, and the left tooth of the narwhale (Figure 14).

We see that these are slight alterations of bilateral symmetry in response to particular constraints rather than true examples of triple polarity. Hands are organs with three polarities but they come in pairs as mirror images, and the pair has one symmetry and two polarities, as is the norm. In the end, the only really good example of triple polarity is the mythological dahu of the mountains of Europe (Figure 15). In the remarkable dahu, loss of symmetry corresponds to the continual need to move horizontally along the steep slopes.

Changing the Scale

Now let us turn to the microscopic scale (0–100 μm), that of unicellular organisms: bacteria, amoebas, euglenas, peridinians, desmids, diatoms. D'Arcy Thompson (1917) noted that this scale represents the frontier of a world with which we have scarcely any experience, that "The preponderant parameters here do not operate at the scale for which we are familiar; we are at the frontier of a world of which we have scarcely any experience, and we must revise all of our concepts." Gravity becomes negligible compared to other forces—surface tension, viscosity, friction, Brownian motion—and vertical polarity simply does not exist (Figure 16).

Figure 15. The mythological European dahu. Above, walking horizontally on a steep slope. Below, an exceptionally well preserved dahu skin at the Regional Museum of Chamonix.

An extremely important factor in the lives of unicellular organisms is that their small dimensions confer on them vast surface areas in relation to their volumes, and this gives them advantages in absorbing solar energy and capturing prey. They are at once a surface and a volume, a plant and an animal. In fact, morphological differences between the two kingdoms are not seen at this level. We understand why unicellular plants and animals, functionally indistinguishable, preceded multicellular organisms evolutionarily. Conquering gravity, they were forced to choose, to be plant or animal.

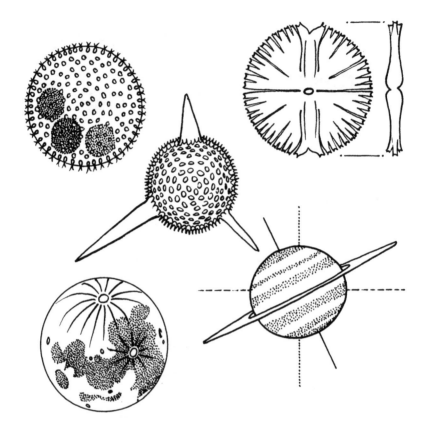

Figure 16. Above, symmetries and polarities at a microscopic scale: the protist *Volvox* (left), the radiolarian *Haliommatina* (center), and the desmid *Micrasterias* (right). Below, comparable symmetries but at an astronomical scale: the Moon and Saturn.

A hiatus of 0.1–1 mm (100–1000 µm) separates the microscopic and macroscopic scales of life. Biologically, this intermediate zone is almost empty, favoring neither vertical polarity nor bilateral symmetry. However, this zone is not completely devoid of life. A singular organism, *Dictyostelium discoideum*, lives as three different forms. In the first form the cells are separate, amoebas about 10 µm long, living primarily on bacteria (Figure 17A). In the second form, in which no food is consumed, the amoebas converge in the thousands to form a multicellular organism (Figure 17B) like a small slug about 4 mm long that moves and leaves a trail of slime behind (Figure 17C). This form, bilateral, anteroposterior, and dorsiventral, is like

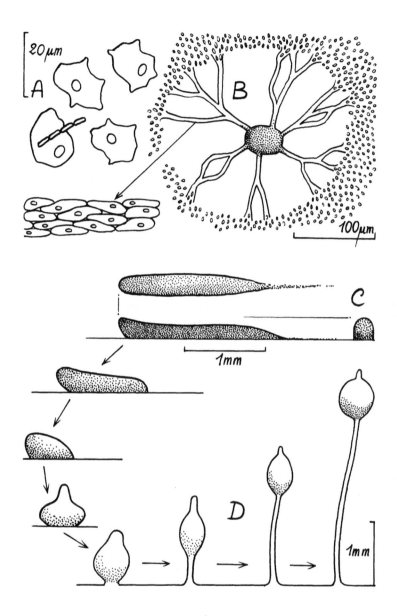

Figure 17. An organism that changes form in crossing the barrier of 100 μm
to 1 mm. The cellular slime mold *Dictyostelium discoideum* has a stage at
which its amoebas dissociate and feed on bacteria (A). If food is lacking, the
amoebas converge into a multicellular organism (B) that becomes mobile,
taking the form of a slug (C) before becoming fixed in place and adopting
a plant-like form (D). In the form of a protist when smaller than 100 μm,
D. discoideum adopts an animal or vegetable form when larger than 1 mm
(Raper 1940, Bonner et al. 1955, Larpent 1970).

that of an animal. After several days of life and movement, the slug stops moving, becomes round, and produces a vertical stem about 5 mm high, sprouting a sporangium at its tip. Radial in symmetry and with vertical polarity, this third stage is like a plant (Figure 17D). What is interesting in this example is that it shows that form depends on scale, and one form is only functional at one scale. At a scale smaller than 100 µm, *Dictyostelium* takes the form of a protist; above 1 mm it takes the form of an animal or plant.

Does this change in form with size occur only in the animal kingdom? "A significant gulf separates unicellular and multicellular organisms. There are no 'meso-organisms'," Grassé (1973) wrote. There are Mesozoa (Margulis and Schwartz 1988), however, whose members develop only a few dozen cells and grow only a few hundred micrometers long. These animals, the simplest of all, live in the kidneys of marine invertebrates—sea worms, starfishes, octopuses—feeding on the urine of these animals. The adults are dorsiventral in symmetry, but the small larvae are spherical.

Do Mesozoa contradict the micro- versus macroscopic? I think rather that their name points to these animals as exceptions in the range of 100 µm to 1 mm, leading me to agree with the hiatus proposed by Grassé. Besides, how astonishing that nature would avoid this intermediate realm where the effects of weight and friction compete, where living creatures are neither large nor small, where they find a niche neither in spherical nor dorsiventral form?

It would be interesting to continue this inquiry into the astronomical dimensions, from 1 km to infinity, where enormous objects outside our familiar space do not have polarity (Figure 16). This can be seen in the forms of galaxies, but that is certainly not my domain.

This brief discussion of relationships between space and form illustrates three important ideas. The first is that vertical polarity is appropriate for macroscopic dimensions. At the extremes, microscopic and astronomical, such polarity does not occur because of the obvious effects of scale. The second idea is that the nature of the object is not important in the form–space relationship. Natural or man-made, living or nonliving, this is of no importance concerning form, symmetry, and polarity. D'Arcy Thompson was right. Notably, organisms are not unique in these ways, and the form–space relationship precedes life. Finally, the third idea is that plants and animals are characterized by forms profoundly different from each other. Schematically, plants have radial symmetry and one polarity, and animals have bilateral symmetry and two polarities (Figure 18). These differences affect everything about them.

One Polarity and Radial Symmetry: Plants

Sessile organisms, animal or plant, have forms with one polarity and radial symmetry: hydra or rain forest tree, sea anemone or wood anemone. There are also a few exceptions among mobile animals; sea urchins, starfishes, sea cucumbers, and comatulid crinoids also have this form, but they are ancestral to sessile groups. Some sessile animals—lamellibranchs, barnacles, acorn shells—do not have this form, but they are related to mobile groups. Form evolves more slowly than behavior. Given the prominence of plants in our landscapes, we shall consider them as the legitimate representatives of this category of form.

The polarity of the vertical axis, from base to tip, lengthening with time, results from the fixation of plants to a substrate. This is reinforced by their sheer weight, "the most permanent force in the world and completely underestimated in plant morphology" (Leistikow 1994). A very detailed radial symmetry, also along this vertical axis, develops in their internal anatomy (Figure 19), giving plants the same silhouette regardless of the angle from which they are observed horizontally. Turn an evergreen oak around: If the environment is uniform, the image remains generally the same (Figure 20). This radial symmetry is the result of the stability in structure that conditions the development and function of the plant. If the radial symmetry of flowers is artificially altered, they attract fewer pollinators (Polak and Trivers 1994).

In overall plant form, polarity and symmetry are expressed in the same vertical axis, and this is of great importance. Why is the right angle rare in nature but ubiquitous in the products of human industry? The best example of a natural right angle, perhaps the only one, is that of a tree trunk with the horizon (Vogel 1988). Certainly, such an arrangement gives a tree its stability by keeping the center of gravity of the crown above the base.

In plant biology, the vertical axis (in contrast to the horizontal) lays out a sort of dominance and precedence at once embracing and ancestral. Certainly, the horizontal produces more advanced prop-

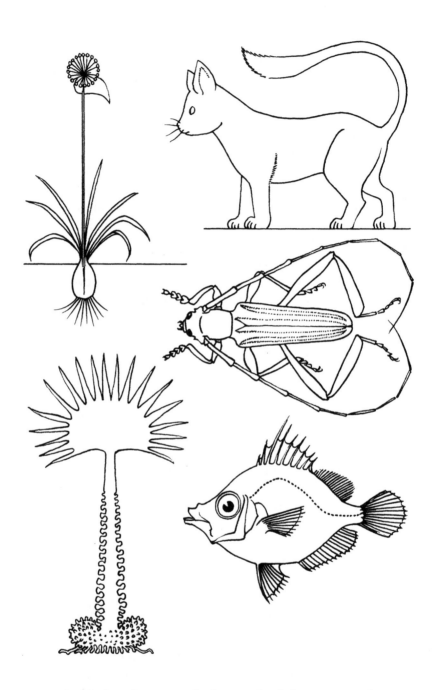

Figure 18. Left, plants have vertical polarity and radial symmetry. Right, animals have dorsiventral polarity and anteroposterior and bilateral symmetry. Below, life in an aquatic environment does not affect the morphological rules, as shown by the marine alga *Saccorhiza bulbosa* and the fish *Capros aper*.

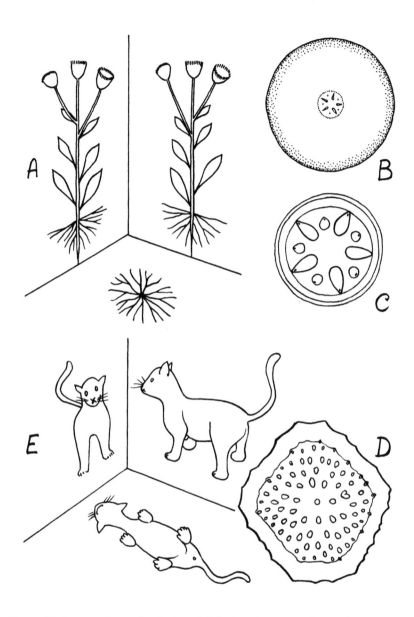

Figure 19. Symmetries and polarities. (A) A plant is projected onto three perpendicular planes; the image is practically the same on the two vertical planes, showing radial symmetry along the vertical axis. (B) Cross section of the branch of a dicot, a fig tree, *Ficus*. (C) Detail of the central cylinder of the fig branch. (D) Cross section of the branch of a monocot, *Ruscus aculeatus* (B–D, Roland and Roland 1995). Notice that radial symmetry is present, even if only approximately, no matter what the scale. (E) An animal is projected onto three perpendicular planes; the images are different on the two vertical planes.

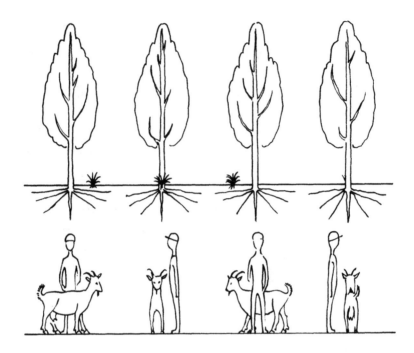

Figure 20. If the environment is homogeneous, a plant presents almost the same image no matter what the angle of observation horizontally. For animals, the image depends on that angle.

erties and functions, and only a little verticality seems needed. The fundamental totipotency of the plant disappears horizontally under the force of gravity. Plants provide many examples demonstrating the preeminence of the vertical over the horizontal. The apical meristem, which produces the vertical axis of a plant, divides into small lateral meristems that later produce leaves. These organs, fundamentally horizontal, are very active but have only a brief life span; the vertical meristem continues to split off new leaves while old ones fall to the ground. Decapitated, the vertical plant axis is rapidly regenerated in most cases by a lateral meristem. In contrast, if a leaf is cut during development, there is no recovery and it remains cut at maturity. A plant such as coffee, cedar, or cocoa produces a mixture of vertical and horizontal axes. The rooted cutting of a vertical axis will produce a normal plant whereas that of a horizontal axis will only produce a horizontally spreading plant, only good for ground cover. The only animals in which the vertical axis is as important as in plants are sessile, for example, corals (Chapter 6).

Two Polarities and Bilateral Symmetry: Animals

Forms with a unique symmetry and two polarities are characteristic of most animals: stag beetles or kangaroos, flatworms or cormorants. Some plants also have a similar form, but the striking numerical superiority in the animal world lets us consider free-living animals as the legitimate representatives of this category of form.

First, the two polarities: One is represented by dorsiventrality, linked to gravity and contact with the substrate. It is manifested by the contrast between the back and the belly, the latter accompanied by limbs. The other polarity is anteroposterior (or more technically, cephalocaudal), linked to mobility. This gives the animal a head, tough and profiled, permitting easy penetration of its milieu, and a tail, whose function is less important since it follows the movement. Animals concentrate their sensory organs at the front with their anus at the rear, because they need to know where they are going and move away from what they leave behind (Gould 1989).

The unique symmetry of animals is bilateral, and the evolutionary emergence of this symmetry is an adaptation particularly well adapted to the complicated interactions of animals with their environments. For example, movement and the reception of signals from the environment through visual, auditory, tactile, and olfactory sensory organs are enhanced by the bilaterally symmetrical organization of the body and its centers of control (Schonen 1989). As an example, if you walk around a goat (Figure 20) the image you see will change—unless the goat, intrigued, decides it is more prudent to face you.

Bilateral symmetry has an important biological significance for animals (Polak and Trivers 1994). Individuals who achieve it best are strongest, find sexual partners more easily, are less susceptible to parasites and disease, etc. In effect, the developmental program of an animal is perfectly symmetrical, but some individuals may not have the ability to complete the program well. The result may be a series of small deficiencies in symmetry, which may be termed fluctuating asymmetries. Such fluctuating asymmetries may give a measure of the fitness of the phenotype. In humans, these asymmetries are particularly strong in individuals whose mothers abused tobacco or alcohol, or those who suffer various mental disorders resulting from chromosomal abnormalities, such as that for Down's syndrome (Polak and Trivers 1994). This is consistent with the view of ancient Greeks, in which the optimal state was seen as perfect symmetry.

And Monsters?

Artistic license is not exempt from the constraints of symmetry. Even monsters such as those from frightening books (Hieronymus Bosch, Francisco de Goya, Gustave Doré), from the fantastic bestiary of Jorge Luis Borges, from comic books, or from horror films and science fiction (the sympathetic E.T.) are most often endowed with bilateral symmetry, dorsiventrality, and a cephalocaudal polarity (Figure 21). Such artists would wish to be free of all rules and unshackled, only wishing to surprise or frighten. As fantastic as they are, however, these creatures docilely conform to the constraints imposed on mobile creatures by space and form. If this conformity were lost, they would no longer be monsters but simply amorphous objects without any significance, frightening no one. The form of monsters is little studied but there are some references (Wyndham 1954, Mezières and Christin 1971, Tardi 1978, Huygen 1979, Hawkins 1990, Gidoin 1991).

Notice that bilaterality, which occurs in the domain of the nonliving as well as the living, concerns only the "external organs, in contact with the atmosphere" (Thom 1988). Pig or dragster, airliner or ladybug, these mobile objects are symmetrical only in regard to their interface with the exterior; in their interiors, whether visceral organs or machinery, symmetry is no longer respected. It is true, as Scania de Schonen (1989) has written, that "functions provided by the heart, liver, or pancreas probably gain nothing from symmetrical disposition of these organs" (Figure 22). This merits some consideration. Why is there symmetry in a number of paired organs, such as lungs, kidneys, testicles, and cerebral hemispheres, whereas many unpaired organs are asymmetrically placed, such as the stomach, large intestine, liver, spleen, pancreas, and heart?

Finally, notice that if most animals are in the category of "one symmetry, two polarities," they are not the only such organisms.

Figure 21. Symmetries and polarities of monsters. (A) The 1880 cyclops of Odilon Redon. (B) The centaur of Ulisse Aldrovandi (1553–1605). (C) The space monster of Colin Hawkins (1990). (D) Gold-covered bronze of Bastet, the cat goddess of ancient Egypt. (E) A 1958 bronze by Zev. (F) An elf of Flagh-Staad Fjord (Gidoin 1991).

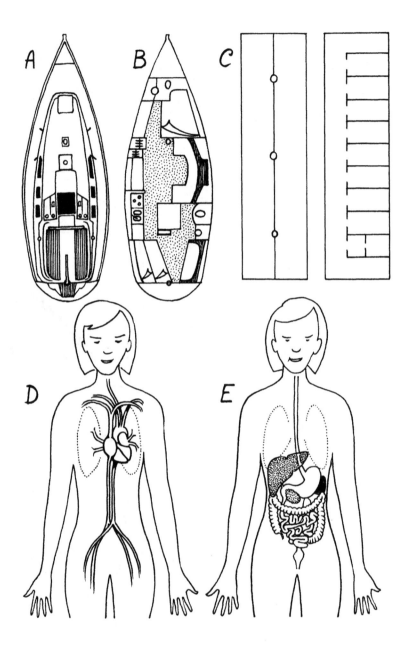

Figure 22. Internal asymmetries. (A) Vehicles, with some exceptions (Figure 13), have symmetrical exteriors; the deck of my sailboat is a good example. (B) In contrast, there is no reason for such symmetry in the interior of the same boat. (C) Agatha Christie's railroad car on the Orient Express also had a symmetrical exterior but was asymmetrical internally. Human beings are externally symmetrical but most of their viscera are positioned asymmetrically, for example, the circulatory system (D) and digestive system (E).

Plants have explored these niches, which for them represent primitive conditions prevailing at the time of the first terrestrial plants (Leistikow 1994). We may believe that these plants were not numerous, that there was little competition for light between them, and that they could thus spread out in the sun. Early land plants with one symmetry and two polarities were not numerous but they do provide more evidence of plant versatility (Figure 23).

How had evolution come to rely on one or the other of two systems as different as "radial symmetry, one polarity" and "bilateral symmetry, two polarities"? This is a legitimate question since the two systems coexisted among plants as well as animals. Even if it is just speculative, it is possible to imagine a means of connection between the two systems that divide the world of multicellular organisms (Figure 24). Changing from the radial system, a reduction in the height of an organism can lead to a point where the relationship between volume and surface becomes propitious for active mobility, the latter imposing bilaterality. Echinoderms demonstrate this type of evolution, from the crinoids, which formed veritable marine prairies of radial and sessile animals during the Silurian, to modern sea urchins, in which we can see bilaterality established along with mobility. Changing from the bilateral system, a modification of ecological conditions accompanied by competition for light would necessitate a return to a vertical axis and radial symmetry. Thus, parting from the ancestral creeping forms, modern terrestrial plants would have appeared (Leistikow 1994). In this view of the evolution of form, and it is speculative, plants and animals evolved in opposite directions, the point of departure for one being the point of arrival for the other.

What Is an Embryo?

In the preceding discussion, the form of organisms was viewed as an established theme, readily observable. But a dynamic process imposes itself on form: embryogenesis, the formation of the embryo, and more generally, morphogenesis. Is the development of form the same in plants and animals?

Embryo is a word used for both kingdoms. It designates the result of morphogenesis in the period from the first division of the egg to the beginning of independent life: hatching, birth, or germination.

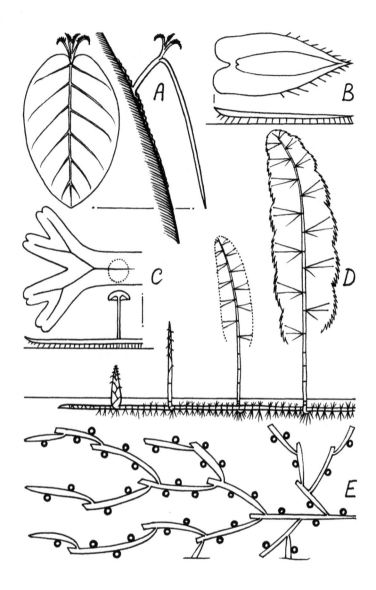

Figure 23. Some examples of plants with bilateral symmetry and two polarities. (A) *Streptocarpus* and *Epithema* are African Gesneriaceae whose structure is established by the development of a cotyledon that grows as long as 30 cm; the other cotyledon disappears. (B) Gametophyte or prothallus of a fern, 20 mm long. (C) *Marchantia,* a liverwort, 30 mm long. (D) *Phyllostachys,* a Japanese bamboo whose erect culms, more than 10 m high, are derived from the branching of subterranean stems. (E) The subterranean stems of *Phyllostachys* also branch horizontally, giving the plant the form of a vast leaf dozens of meters long, shown here only in part. Circles indicate the positions of the erect culms.

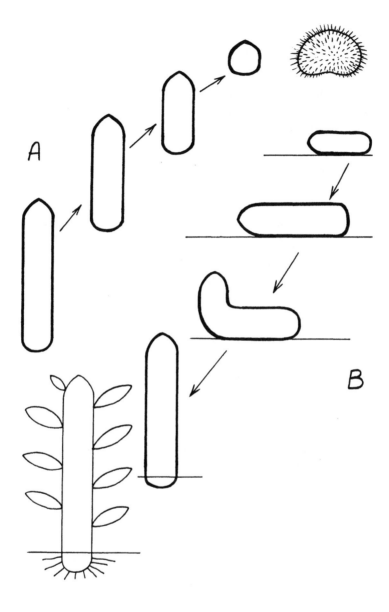

Figure 24. Shifts between systems that are radially symmetrical with one polarity or bilaterally symmetrical with two polarities. (A) Changing from a radially symmetrical system, a decrease in height leads the organism to the point where the relationship between volume and surface area favors mobility, which imposes bilateral symmetry, as observed in sea urchins. (B) Beginning with a bilaterally symmetrical system, competition for light has necessitated vertical growth and thus the adoption of radial symmetry; terrestrial plants could have had such an origin (Leistikow 1994).

During this period, maternal energy supports the embryo, which is by definition a parasite. Is the transition from single cell to organism, the essence of embryogenesis, the same in plants and animals? At the beginning, the two processes seem identical in two well-studied examples: *Capsella*, shepherd's purse, the tender plant I observed on the window ledge in Paris, and *Xenopus*, a sort of poor example of a toad.

After fertilization of the female gamete by the male gamete, an egg is formed. This is the first cell of the embryo, the first cell of the future plant or animal. This cell-egg, in *Capsella* as well as *Xenopus*, displays an upper side and a lower side from the very beginning; each is polarized (Figure 25). In the egg of *Capsella*, nicely elongated, the nucleus is located above, and the base is marked by a vacuole. The *Capsella* egg already has the radial symmetry and base–tip polarity that will characterize the plant throughout its life. The direction of this polarity is produced by the meeting of the two gametes. The origin of the future root is determined by the point of entry of the sperm nucleus into the female gamete (Figure 26).

When it divides for the first time, the egg of *Xenopus* is a sphere but its contents are not symmetrical. An animal pole, above, is distinguishable from a vegetal pole, below, which is loaded with nutrients. Note the droll vocabulary adopted by the zoologists, which expresses a tranquil conviction that animals are more noble and alive than plants. The principal axes of the body of *Xenopus* are defined just as early as those of *Capsella*. The animal pole becomes the back, and the vegetal pole, the belly. The dorsiventral axis is already in place before the first cell division. The entry point of the spermatozoid determines the position of the head, while the position directly opposite is the gray crescent, and bilaterality is established for the rest of the animal's life (Figure 25).

Homeotic Genes

When one part of an organism resembles another in the course of development, the process explaining this pattern has been defined as homeosis. A homeotic mechanism expresses a character at site A in an organism, then expresses it as another character at site B. Examples of homeosis from both plants and animals are grouped together in Figure 27 because the mechanism occurs in both kingdoms.

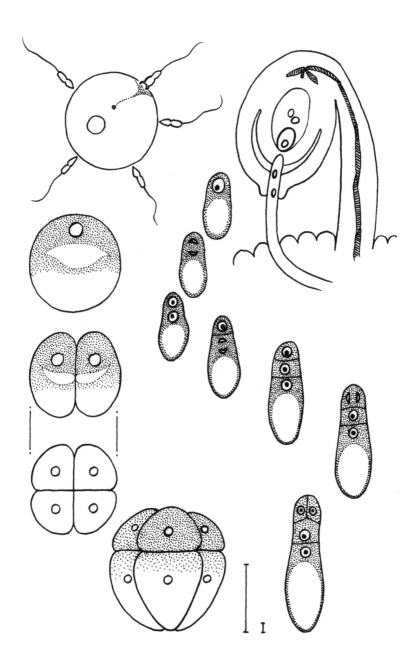

Figure 25. Early stages in the development of the animal *Xenopus* (left) and the plant *Capsella* (right). In both, the egg cell already has the polarity and symmetry of the adult, profoundly affecting the first stages of embryogenesis. Below, side by side, a *Xenopus* embryo with eight cells and a *Capsella* embryo with four cells. Between them, scale bars for the two embryos: 1000 μm for the animal, 60 μm for the plant (Souèges 1919, Alberts et al. 1986).

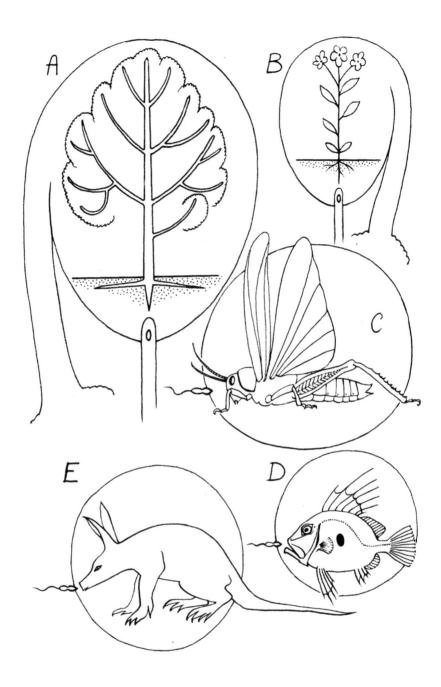

Figure 26. Orientation of organisms and the entry point of the male gamete into the female. In plants, whether a tree (A) or an herb (B), the entry point of the pollen tube into the embryo sac determines, in most cases, the point of the first root. In animals, whether insect (C), fish (D), or mammal (E), the entry point of the spermatozoid into the egg determines the location of the head.

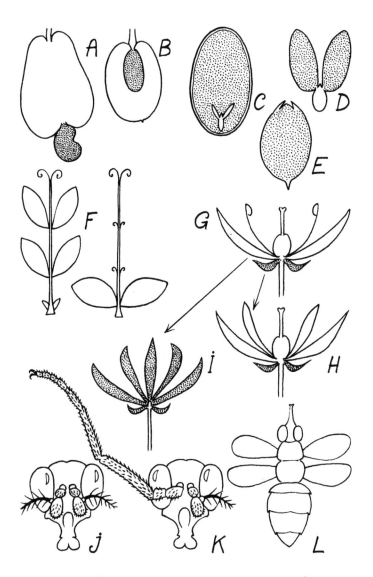

Figure 27. Examples of homeotic modifications. In the cashew (A, *Anacardium occidentale*), the fleshy peduncle replaces the fleshy mesocarp of a plum (B, *Prunus*). Seeds can accumulate reserves in three ways: in the albumen (C), cotyledons (D), or the axis of the embryo (E). (F) Two different species of *Lathyrus;* on the right, leaflets have disappeared and have been replaced by stipules. From a normal, perfect flower (G), a doubled flower is produced in which stamens have been transformed into petals (H). (I) In the *leafy* mutant, all floral parts take on the appearance of sepals. (J) A fly head with two antennae. (K) In the *antennapedia* mutant, one antenna is replaced by a leg. (L) A fly with the *bithorax* mutation (Leavitt 1909, Corner 1958, Le Guyader 1994, Meyerowitz 1994).

In the 1990s, analysis of mutations provided evidence of many homeotic genes responsible for development in different parts of organisms. Although their effects are often monstrous when the genes mutate, these homeotic genes are responsible for early differentiation of the embryo. This differentiation (Figure 28) occurs along the vertical axis in a plant (Jurgens 1992) and along the anteroposterior axis in an animal (Le Guyader 1994). The position of genes in the segments of DNA controlling form in the fruit fly *(Drosophila)* corresponds to the position along the body of the fly (Figure 28C). This was an unexpected resurfacing of the notion of the animalcule, which until the 19th century was thought by embryologists to prefigure the form of the adult in the egg.

Molecular embryology, as recent as it seems, has some old roots. These are found in the "transference of function" of Corner (1958) and even in the "morphic translocations" of Leavitt (1909). For those who passionately follow progress in molecular embryology, the difference in approach to research on homeotic genes in plants and animals is regrettable. For animals, the complete embryogenesis of many groups, from nematodes to mammals, is well established. For plants, most research is on genes that control flower development, including sexual organs and accessory structures, of just a few plants (Coen and Meyerowitz 1991, Schmidt et al. 1993, Ma 1994, Meyerowitz 1994): *Arabidopsis, Antirrhinum,* tobacco *(Nicotiana),* and maize *(Zea).* This difference has yielded quite different results for plants compared to animals. It would be more fruitful to compare similar phenomena in the two kingdoms to understand developmental processes. Jurgens (1992) made such a comparison and noticed a hidden similarity among the homeotic gene families of plants and animals, not only among those of diverse groups of animals, concluding that there are common principles in the construction of multicellular organisms, plant or animal. Unity in the development of living form is, in my view, where molecular embryology can best lead us.

Animal Eggs, Plant Eggs

Revisiting symmetry and polarity, it is interesting that the major axes, which constrain organisms for their entire lives, are expressed at the earliest stages of development, beginning with fertilization

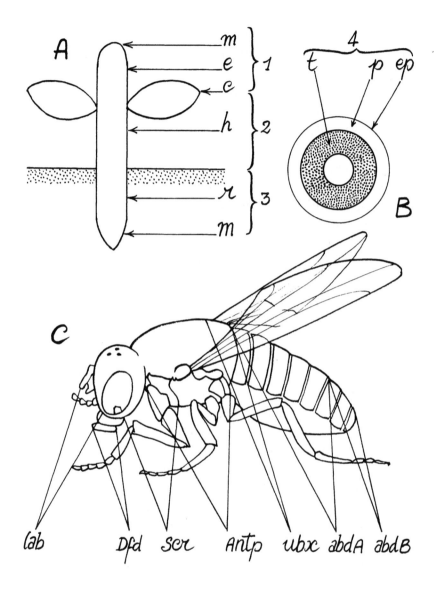

Figure 28. Developmental regions of embryos. (A) In plants, after the very early stage of eight cells, three groups of genes form three regions lengthwise: 1–3 (m, meristems; e, epicotyl; c, cotyledons; h, hypocotyl; r, root). (B) A fourth genetic mechanism establishes anatomical zonation: ep, epidermis; p, parenchyma; t, vascular tissues (Jurgens 1992). (C) In *Drosophila*, the positions of the genes that control developmental regions of the embryo correspond to the anteroposterior positions of the products of these genes in the body of the fly. Thus *lab* is responsible for the antennae, *AbdB* for the terminal portion of the abdomen (Le Guyader 1994).

and even before the egg has divided. After cells begin dividing, the eggs of *Xenopus* and *Capsella* develop by different mechanisms. The egg of *Xenopus* (Figure 25) divides twice, followed by vertical partitioning or, more precisely, by mitosis with a horizontal spindle, producing four cells practically identical in size. Each cell has an animal pole and a vegetal pole. If the four cells are experimentally separated at this stage, they produce four identical and perfectly viable individuals: clones. Only later, with the third cycle of cell division, are cells of different sizes produced, four small ones above and four large ones, enriched with food reserves, below.

On the contrary, the egg of *Capsella* divides by horizontal partitioning, by mitosis with a vertical spindle, and the two resulting cells are different from the very beginning. The cell above has a dense cytoplasm, and the cell below is marked by a large vacuole (Figure 25).

Other differences appear. In *Xenopus*, the first 10 cycles of cell division are synchronous. In an embryo with more than 1000 cells, some relatively small and others large and full of food reserves, all divide at the same time. It is only in an embryo with more than 2000 cells that divisions cease to be synchronous. In *Capsella*, cell divisions are never synchronous. From the third mitosis (Figure 25), the cell at the bottom divides whereas the one above divides later. This asynchrony seems to me to be the first evidence of the absence of centralized control, an absence that is characteristic of plants.

Differences increase with further development. The *Xenopus* embryo first passes through the stage of a small spherical mass of cells, which is curiously baptized with a botanical name, morula, little mulberry (Figure 29). Next, it transforms into a blastula, hollowing out an internal cavity, the blastocoel. The gastrula stage is critical. The embryo forms a cavity by invagination of the wall without increasing in volume, creating an internal cavity that opens to the exterior through a blastopore. The embryo is thus a structure twice hollowed out, in which contact with the external milieu is now provided by internal surfaces. These latter surfaces, which increase during the rest of the developmental period, become the digestive surfaces across which food is assimilated by the animal.

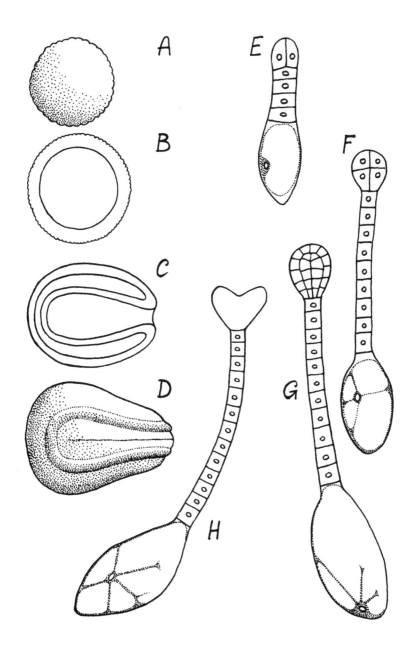

Figure 29. Embryology of the animal *Xenopus* and the plant *Capsella*. In the animal, the embryo passes through the stages of morula (A, seen externally), blastula (B, seen in cross section to show the blastocoel), gastrula (C, seen in a section passing through the blastopore), and neurula (D, seen externally to show the position of the nervous system; Alberts et al. 1983). (E–H) The plant embryo, with suspensor and basal vesicle; cotyledons begin to appear at H (Souèges 1919).

Animals Are Strange

It seems logical that the blastopore, the opening connecting the digestive tract with the exterior, would become the mouth. This is true for invertebrate animals: snails, earthworms, crabs, insects, scorpions. In vertebrates, however, such as *Xenopus* or humans, the blastopore becomes the anus, and the mouth only appears later at a point diametrically opposite. Thus, when you encounter a loud-mouth, tell him that anyone who produces an anus before a mouth, and thus farts before being able to speak, should be more modest. Yes, animals are strange.

At a later developmental stage, that of the neurula, tissues begin to form that will become the animal's nervous system. Such events are of no concern to plants. From the beginning, plant embryos unfold quite differently, and the *Capsella* embryo shows this very well (Figure 29). It never folds in on itself, unlike animal embryos. It grows while remaining solid and convex, and its surface is the one through which solar energy and water will enter the plant.

I will not linger too long on the well-studied mechanisms that produce completely developed plant and animal embryos, but I would like to show that we stand before two different worlds. Animal embryogenesis is a complex phenomenon in which numerous mechanisms operate concurrently, relying on each other through a highly integrated process. The result is at the same time precise and remarkably detailed. Each cell of an animal embryo functions and differentiates according to the particular information it receives. One source of information is the cell's location in the layers of the gastrula: A cell in the endoderm can only participate in the production of the digestive tract, whereas a cell in the ectoderm can only produce a neuron.

Animal embryos develop particularly through cellular differentiation. Some pathways used for control of the mechanisms, such as gene regulation, remain poorly understood. Others, such as hormonal action, are studied experimentally. Animal embryos also use movement for development: Cells are mobile and they migrate; tissues fold into grooves or tubes that later develop into organs. In mammalian embryos, a good example of such cellular mobility is the early migration of germ cells, whose descendants will produce gametes, toward the genital crest, which becomes the gonads. This development results in the phenomenon of induction, an effect of interactions between territories, or between cells. Cell division can occur in any part of a structure, and apoptosis, programmed cell death, is frequently employed. At the end of a complex and precise succession of events, the animal embryo becomes a miniature but complete individual, at which time birth or hatching becomes possible. The succession of embryonic transformations seems completely predetermined. We know in advance that the young puppy will have two ears and four paws.

Even if some cell types are absent at birth, such as gametes, the animal embryo clearly resembles the future adult, apart from size (Figure 30). All the internal organs are present and functioning. Animals that develop through metamorphosis—crustaceans, insects, amphibians, some fishes—only develop approximately by this plan.

Plant embryogenesis is simpler. It uses a process that is much less integrated and that does not rely on hormones. Plants produce growth-regulating chemicals, but they play a less visible role in plants than hormones do in animals. Other differences are discussed later.

Hormones

At the beginning of the 20th century, Ernest Starling (1905), a British physiologist working on vertebrate digestion, discovered that when an animal eats, its intestinal wall secretes a chemical that enters into the blood circulation and stimulates pancreatic function. Comprehending that this substance (secretin) was not an isolated example, Starling coined the term hormone for a chemical compound that functions in the henceforth classical sequence: produced by organ A, transported by blood vessels, finally acting specifically on organ B at a precise concentration. The study of animal hormones was born.

At about the same time, plant physiologists were trying to understand the results of experiments conducted half a century earlier by Charles Darwin and his son Francis, who studied the growth of the tip of a germinating barley *(Hordeum)* embryo toward light, a phenomenon that became known as phototropism (Figure 31). This tip, or coleoptile, grows toward light (A). The Darwins had shown that placing an opaque cup over the tip made the coleoptile insensitive to

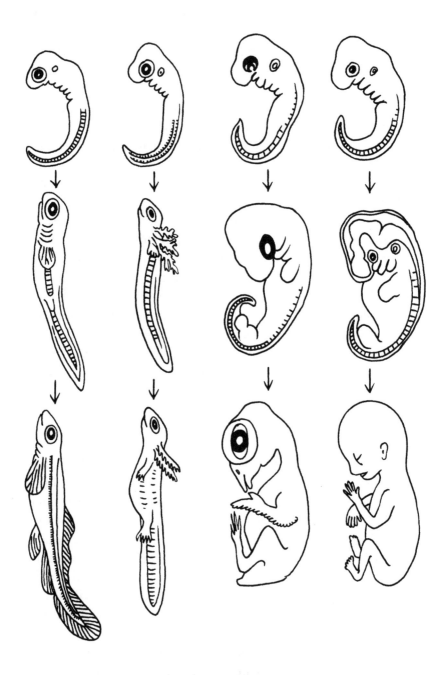

Figure 30. The animal embryo is a reduced model of the adult. Whether fish, salamander, bird, or human (left to right), the early stages (above) resemble each other, and later stages (below) resemble the adults (Haeckel 1897).

light (B). Cutting off the tip also made the coleoptile insensitive (C). Other scientists then demonstrated that placing a small cube of gelatin between the cut tip and the rest of the coleoptile restored phototropism (D). If a chemical compound moderated the phototropism, it was capable of crossing the gelatin barrier. In darkness, it was possible to place the cut tip on one side of the cut coleoptile and obtain the same response (E). The cut tip's influence could also be substituted with a cube of gelatin into which the chemical had diffused from the tip (F).

Plant physiologists are notorious for adopting concepts from outside botany and apart from the plant itself. This psychological trait is evident in the strong influence of Starling's work and their adoption of his concept of the hormone (Weyers 1984). This allegiance to animal and human physiology at first yielded, at least seemingly, excellent results. The term phytohormone was coined and immediately adopted.

Certainly, the transfer of animal endocrinology to plants was compromised by one difficulty: Plants do not have a circulatory system. Admittedly, transport by sap could substitute for this missing element. Plant phototropism received an apparently satisfactory explanation in the theory of hormonal mechanisms derived from the study of vertebrates. Synthesis, transport, and action, the classical sequence in animals, seemed to be discoverable in plants. The organ of production, synthesizing the phytohormone, was in the tip of the coleoptile; the organ of reception was the zone of cellular elongation, notably behind the tip. The sap assured transport, and it only remained to isolate and identify the molecule.

In 1937, Frits Went and Kenneth Thimann showed that indoleacetic acid, or auxin, could replace the tip and induce curvature of the coleoptile in the absence of light. Isolated in various parts of the plant, auxin was thus the first phytohormone to be identified, but many others were discovered later. Born 30 years after its big sister, phytohormone physiology progressed during the next half century, while agriculture began to come up with applications: killing weeds, accelerating crop growth, promoting flowering, controlling the timing and quality of fruit production. All this was enormously profitable, and further research was well supported. By 1980, plant hormones were a major area of research in plant physiology (Champagnat 1987).

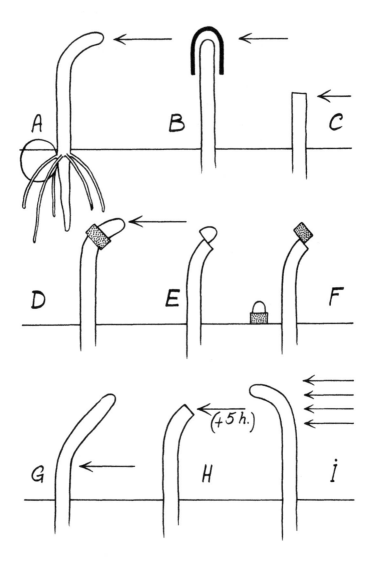

Figure 31. Phototropism. (A) The coleoptile of a grass is attracted to light during its growth. (B) Cowled, the tip of the coleoptile becomes insensitive to light. (C) Decapitated, the coleoptile becomes equally insensitive to light. (D) Intercalation of a layer of gelatin restores phototropism. (E) Placing the tip off-center makes the coleoptile curve in the absence of light. (F) The off-center tip can be replaced by a piece of gelatin on which the tip had rested. (G) A response to light is obtained by illuminating the base of the coleoptile. (H) The decapitated coleoptile becomes sensitive to light again after 5 hours. (I) The coleoptile turns away from illumination that is too strong. Results A–F are compatible with the hormonal hypothesis; results G–I are not (Weyers 1984).

Nevertheless, the first cracks in the edifice appeared before 1980. The phloem did not provide the predicted means of transport, and the term hormone had been criticized and replaced by growth substance, then by growth regulator. But the sense that these concepts were inadequate remained from observation that growth was not always the result, as in the maturation of fruits. A physiologist at the University of Edinburgh, Anthony Trewavas, shattered the traditional understanding. His critique of hormonal theory was severe, but it is also interesting in the context of comparing plants and animals.

Trewavas (1981) first attacked the hormonal interpretation of phototropism. He remembered that 10 years earlier some doubt had been cast on the validity of the experiments: The gelatin had not been sterilized and contained bacteria capable of producing auxin (Libbert and Manteuffel 1970). Trewavas argued that there was no correlation between the concentration of auxin in the zone of elongation and the extent of that elongation. Since animals respond to precise doses of hormones, might plants be capable of responding to a large range of concentrations? Three experiments embarrassed the advocates of the hormone orthodoxy. Phototropic responses were observed by illuminating the base of the coleoptile (Figure 31G). A decapitated coleoptile, at first insensitive to light (C), became sensitive again after several hours (H). Finally, the coleoptile turned away from light of very high intensity (I). We have systematically disproved all the predictions of the hormonal mechanism, said Trewavas.

The Action of Light Remains a Mystery

Trewavas asked, Do plants require hormonal control? He proposed an alternative hypothesis in which light would be the direct cause of the bending of the coleoptile, by controlling the levels of nutrients in the elongating tissues. This hypothesis was supported by the evidence, without which we would have pretended to understand phototropism perfectly. Weyers (1984) rallied to his support, recognizing that the action of light had remained obscure.

What is important here is to recognize that an animal mechanism does not work in plants. In effect, it was through comparison of the two kingdoms that Trewavas found his critical ideas. Since plants and animals diverged from each other more than 550 million years ago from unicellular ancestors, is there any reason to suppose that in multicellular organization the means of communication between cells should be the same? His comparison of the two kingdoms strengthened his conviction.

Animals reveal significant differences between tissues composing their numerous organs. This tissue specialization within an individual allows the endocrine glands to coexist with groups of target cells, so the system is capable of delivering messages to distant targets. It is this way, wrote Trewavas (1981), that animals establish their principal characteristics: determinate development of structure and centralized control of function.

There is nothing comparable in plants, in which a continuum of identical tissues would impede the recognition of permanent donor and receptor organs. According to its stage of development, each plant cell is capable of producing or using all the molecules considered to be phytohormones (Champagnat 1987). Even if these substances, applied artificially, find important applications in agriculture, even if their presence in plant tissues has been demonstrated, they are not hormones. They do not have a particular place of production, a means of transport, or a specificity of action.

Determinate development and centralized control are processes unknown to plants. Plants do not produce hormones because they do not need them. The hormone hypothesis is useless because competition between meristems for resources—space, light, carbon fixed by photosynthesis, minerals in solution—produces situations in which change in part of a plant affects other parts at considerable distance. This affair of hormones in the history of plant physiology will eventually be recognized as an error in thinking, resulting from the application of a zoological concept to plants. It will take some time for this consensus to be reached; we still see the term plant hormone used in the literature, as in writing on apical dominance (Cline 1997).

The Miniature Model Versus the Sample

It is time to return to *Capsella* and *Xenopus* embryos to see a profound difference in the modes of their development, when the young organisms become autonomous. The newly hatched *Xenopus* is a miniature of the adult. At germination, the *Capsella* embryo is a sampling of the adult, very much like the trial samples of a perfumer or a cloth merchant.

In a growing *Capsella* embryo, most cells soon lose their capacity to divide, and mitosis becomes localized in two particular zones, the meristems (Figure 32). The cells that stop dividing expand in volume, and embryonic growth is primarily the swelling of these cells. In contrast, meristematic cells are not numerous and hardly differentiated. The young plant only vaguely resembles the future adult. The structures of the two do not seem to be the same. The *Capsella* embryo consists of a single root and a stem bearing two leaves (Figure 32). It is thus only a sample of the organs that appear later during development, which multiply into many copies. In trees, such multiplication is not by the hundreds, or thousands, but hundreds of thousands. It is impossible to confuse embryogenesis and development in plants, unlike in animals. Only when embryogenesis is completed can development begin.

Another difference between the two kingdoms appears at this point. Animals possess numerous organs, each with no or few copies —one brain, generally two eyes, one heart, one sex organ, two kidneys, one liver, one mouth, etc.—whereas plants are composed of the three organs understood by Goethe: root, stem, leaf. Even a flower, a fruit, or a seed is reducible to these three organs. Even if it is satisfactory in most cases, this classification of plant organs is certainly too strict. Intermediate forms develop between the three, and these may precede the evolution of the normal organs (Sattler and Jeune 1992). That there is a continuum of organs (root–stem–leaf), to which can be added trichomes (hairs), confirms the concept of the totipotency of cells and tissues so characteristic of plants. Comparing the embryo of *Xenopus* near birth with that of *Capsella* prior to germination, we find them too different to allow subsequent development to be identical. The future of a miniature model and that of a sample can scarcely be the same.

Closed and Open Development

The animal embryo, at least partially constrained in its independence after birth, only needs to grow to achieve adult size, all the while conserving the same structure. The plant embryo, also partially con-

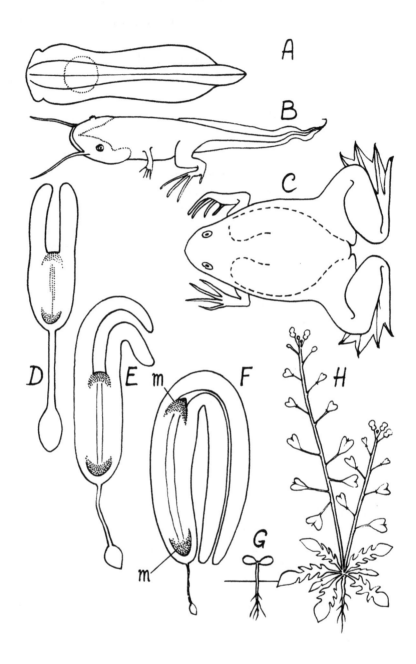

Figure 32. Later stages in the embryology of the animal *Xenopus* and the plant *Capsella*. (A) *Xenopus* embryo budding a tail (dorsal view). (B) Tadpole. (C) Adult *X. laevis*. (D) *Capsella* cotyledons become clearly visible, then curve (E), allowing the seed to adopt a rounded form (F); the suspensor disappears (m, meristems; Roland and Roland 1995). (G) Germination. (H) *Capsella bursa-pastoris* beginning to flower and fruit.

strained after germination, finds itself confronted with a more difficult task, for practically everything remains to be developed. Its rudimentary structure must be elaborated into the future plant.

In animals, the embryonic stage is gradually blurred during the juvenile phase, even if tissues persist that allow for cellular renewal. Embryogenesis is limited to a short period at the beginning of life. Animal development is thus termed determinate, defined (Margara 1982), or closed (Hagemann 1982). Take the example of the human being; after an embryonic period of 9 months, it attains adult size by 20 years, which it will maintain until death. All animals are the same with the exception of a few groups: fishes, reptiles, marsupials. In these animals, growth continues slowly throughout life.

In contrast, plants have indeterminate, undefined (Margara 1982), or open (Hagemann 1982) development. They continue growing their entire lives, increasing in size and producing new organs: leaves, stems, roots. This development can be interrupted by a combination of unfavorable external conditions or competition between organs, but only temporarily.

We are used to this odd arrangement. In plants, contrary to animals, embryonic development is not confined to a short period at the beginning of their lives. Their clusters of embryonic cells, meristems, are not found in animals and allow plants to grow throughout their lives, giving them an indeterminate embryogenesis. It is a poor translation of reality to say that plants *can* grow indefinitely; they *must* grow, to the extent that they die if so impeded. The life of a plant is mingled with its growth. A steel band around the trunk of a plane tree stops its growth in diameter, then the tree dies. Such a band around my daughter's neck makes her a stylish Parisian (but a simple necklace would be nicer).

In plants, the stages of growth successively accumulate, whereas in animals they succeed and exclude each other with rare exception, such as the scales of fishes and tortoises, and the shells of mollusks. The base of a large tree contains the remnants of its youth (Figure 33). The tree has grown by accumulation, since its cell walls are strong and rigid. On the contrary, in the autopsy of a human adult nothing is found of its youth except a few specialized cell types, representing a juvenility spared by the passage of time.

To summarize: From the beginning of embryogenesis of *Xenopus*, everything converges to ensure that it becomes a hollow volume in

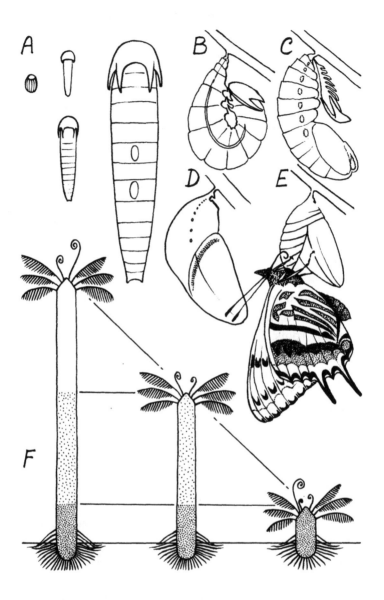

Figure 33. Growth by substitution in animals, and by accumulation in plants. (A) The egg of the butterfly *Charaxes jasius* and its development into a caterpillar. (B) The caterpillar fixes itself in place and adopts a characteristically folded position. (C) Formation of the chrysalis or pupa, only the dried skin of the caterpillar remaining. (D) The burnished chrysalis several hours before opening. (E) Hatching (as documented by J.-L. Bompar). (F) Three stages in the growth of the tree fern *Cyathea camerooniana*. Successive stages accumulate over time, the conductive tissues at the base of the trunk remaining functional as the tree fern grows.

which food and water enter through the mouth. Internal surfaces tend to be maximized for assimilation of food energy into the organism. On the contrary, the external surface tends to be minimized for three reasons: A large surface tends to increase the loss of energy, increase exposure to predators, and limit mobility. *Xenopus* is a good model for animals; animals are mobile volumes.

Capsella is a good model for plants, providing an instructive comparison with *Xenopus*. Plants are structures in which the external aerial surface ensures both the capture of solar energy and its assimilation by the organism. Since this energy flux is modest, its capture must be efficient, and from this fact the external surface must be maximized. Being fixed in place and with open development, the plant can produce enormous surfaces thanks to its meristems. Being fixed in place also facilitates the uptake of water, for its scarcity in the soil requires development of a gigantic subterranean surface. Plants are essentially surfaces of exchange, and they also grow to much greater linear dimensions than animals. We can now understand why the largest living creatures are plants. Several species of trees are taller than 100 m, and some lianas 300 m long. It is possible but remains to be verified that certain lianas may attain lengths of 1 km (*Entada* in Africa, rattans in the Asian Tropics).

Stories of Trees

Another difference, that of the central control, separates plants from animals. No one knows this better than Michel Luneau (1994), who knows how to make the trees speak: For us, say the trees, "all is connected so that there is no need for any particular centralization. Our internal organization recognizes neither God nor a master. It is a free association of elements of different and complementary organs. These obey nobody but themselves and ask of their followers a simple and essential agreement: growth. Each organ is free in the means by which it attains this growth. To each according to its inspiration. . . ."

"This independence has a corollary, that each takes all the resources it can. Our organization requires of its members that they conduct their personal affairs with all the drama they can. If all were compassionate, the safety of the community would not be as assured. These simple rules of life, flexible, liberal, democratic, pluralist, responsible . . . are the opposite of those of the other kingdom, especially in the human body. Let's speak without equivocation: It is a Jacobin God who has created humans, a ponderous God, jealous, authoritarian, cop-like, fierce, holding tightly onto his power, zealously intervening in his omnipotence. His motto: No central control, no security. His moral, except for some inferior, semiautomatic details to simplify existence, to collect and gather waste: Everything thought, spoken, acted, all projects, all plans or programs, must obtain, at the very first, the permission of the brain.

"The world of flesh is a world in training, the world of a child who asks his father's forgiveness for rowdiness, disobedience, whimsy, without feeling guilty. If he runs away, it is never very far or for very long. He stops, repentant, returns to behaving normally, and cortical control is reestablished, which is as much to say that moral order is imposed by the cranium, the crenulated bunker of the neocortex. In humans, an honest organ acts not at its pleasure; it is the brain that does so. 'Fetch!' And the good dog brings it. It has no wants, this organ, it is only a means, an instrument. It is a eunuch, a slave. Or, mysteriously, it gives exactly the opposite impression: to it, freedom; to us, servitude. Take an arm and a branch, so similar in popular imagination. Compare them. The former moves as it wants, the hand touches, holds, clenches, caresses, as it desires. The latter is a prisoner of the crown, depending on the wind for movement; it seems to float, to compose itself nonchalantly, in space. When we look more closely, we see that the arm is only a tout; it has the freedom of a hound responding to the blow of a whistle. On the contrary, from the moment the bud has the opportunity, the branch plays and disports itself as it will. It will be more or less strong, more or less straight and long, accordingly. We know that, living the way we do, we are deprived of mobility and speech, making communication between humans and us so difficult."

This poetic license captures comparative biology with exceptional lucidity. Another poet, Francis Ponge (1942), remarked that plants, contrary to animals, have no vital organs: "No place in their body so sensitive, that pierced it would kill them." What would biology be without poets?

Fixed but Not Immobile

In *Xenopus* and *Capsella*, the relationship to space has been trans-lated into opposing options. However, the situation is less simple than it seems. Although there is plenty of evidence that plants do not move, there are some exceptions (Figure 34) though not so many to cast doubt on the truth that plants are fixed in place. When I return to the Luxembourg Garden, near the statue of José Maria de Heredia, I am certain to find the linden tree under which my mother brought me to play during my infancy. Enraptured, we heard the bees from a nearby hive. The bees are probably no longer there, but the linden is, insensitive to the passing years but truly sensitive to the changing seasons, more beautiful than ever.

Fixed does not mean *immobile*. That is a preconception I wish to demolish with vigor. We commonly think that plants do not move, and the immobility we confer on them contributes to our concep-tion of plants as inferior forms of life. This is an extremely deluded impression on the part of us animals. If they were mobile, we would be obliged to ignore their verticality, their silence, their obstinacy in resisting our concepts of individuality, species, etc. For us who are mobile, this apparent immobility makes us doubt that plants are really alive.

This is a mistake because plants are doubly mobile. First, one kind of movement is passive, a result of the force of wind. This makes the branches of an oak swing to and fro, and throws flowers into the air, like pretty white butterflies the blossoms of *Campanula* or *Gaura*. If the wind blows, a savanna or a field of grain is as full of movement as the surface of the ocean. This is nothing compared to the second kind of mobility in plants. They sneak away in front of our eyes, but at a tempo with which we are not familiar. A zoologist, my friend Mark Moffett (1995), has shown the mobility of plants as they grow. I would not wish to plagiarize him.

Imagine that we are in a tropical rain forest: the cool dampness of the understory, the still air between the mossy bases of trees, the smell of humus, the sunlight as if in an aquarium, the resonant silence. Shafts of light reveal the openings left by the fall of great trees, and in these gaps, at our height, it is easy to observe the rapid growth of innumerable plants. What moves in this landscape? Ani-mals, certainly. Wax bees hum around our ears, a morpho butterfly

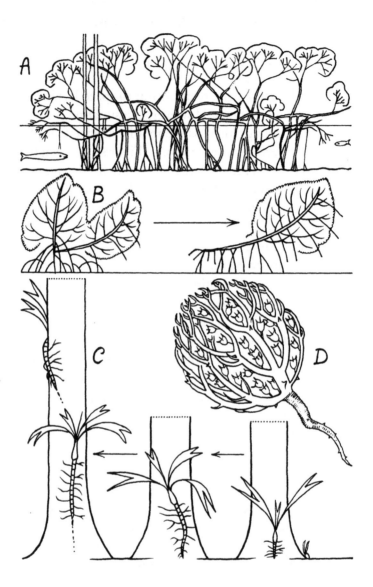

Figure 34. Some mobile plants. (A) *Ficus trigona,* 15 m high, a banyan fig in the flood zone of the Amazon. The site of germination, not detectable, has scarcely any relationship to the actual position of the tree (Sandrine Lamotte, personal communication). (B) A branch of red mangrove (*Rhizophora*) isolated from the original tree moves by growing at one end, dying at the other. (C) The concept of the mobile plant is from Oldeman (1974). A Guianan liana, *Carludovica,* is an example of a plant with vertical movement. (D) The rose of Jericho or resurrection plant (*Anastatica heirochuntica*) rolls like a ball when its fruits are ripe and thus is blown long distances by the wind; it lives in dry regions from Morocco to Iran.

glides by. On the ground, leeches have found us, they hesitate, stand erect on their swollen feet, and begin to crawl toward us. With a little patience we see gibbons pass by above. All the while, the plants seem completely still. However, . . .

The Time Scale for Plants

Concentrating, we see the spiral growth of a vigorous liana, whose speed is like that of the large hand of a clock. Multiply this speed by 100. Thus a minute would correspond to a little less than 2 hours. Animals are now too quick to be seen clearly, but plant movements (in reality, their growth) become obvious. Young branches press toward the sky; young leaves open; the vines circle around; and roots of the strangler fig extend toward the earth. All seems peaceful.

Multiply the progress of time again by 100. A minute of observation now corresponds to about 7 days of real time. The transformation is spectacular. Animals have practically disappeared, only visible in the briefest flashes. Now the movement comes from the plants, from their growth. We easily observe the vigor with which they launch themselves toward the light above the canopy, and we notice competition between them. We see the roots of a strangler fig weld together in a network, slowly closing around the supporting tree. Although growth becomes a source of majestic movement, flowers and fruits develop too quickly to be perceived other than as glimpses of bright color; the time scale of plant sexuality is now that of animals.

Multiply by still another 100, and now a minute of observation covers almost 2 years. Animals have completely disappeared, erased by their mobility. As for the movement of plants, they remain fairly calm in the shade of the understory. This majestic calm is lost in the well-lighted upper strata, where things are more frenetic. The lianas fight with each other in a sort of fierce swarming, then head toward the open canopy like arrows. The sudden skyward launching of leader branches of the great emergent trees corresponds to the hasty and inexorable entombment of the supporting trunk under the network of roots of a strangler fig.

Another acceleration and a minute becomes nearly 200 years. Now, plant movements are too rapid to be seen clearly, but we observe the ecology of the forest in action. The fig has strangled its support and collapsed. Above everything, young trees reach the canopy, exploding into crowns, then collapsing, forming gaps rapidly invaded by pioneer trees and lianas; in 3 minutes the gap is sealed. Other trees fall, forming a mosaic structure in the forest (Moffett 1995).

I hope you are convinced that plants are not immobile, but rather that their movement is not seen in the scale of human time. In their own time, they never stop moving, and we disappear, erased by our own mobility! Remember that the movement of plants is essentially growth. To say that plants are immobile results from an anthropomorphism that impedes our seeing beyond our own time scale. It is as stupid as the history of aphids: In my memory, says the aphid, no one has ever seen a gardener die. Everyone knows that gardeners are immortal.

Movement and Growth

Plant movement does not imply that they travel. Movement and growth are two different phenomena even if they find themselves confounded by expressions such as scrambling plant, creeper, climber, or getting an early start in spring. To avoid this confusion, an illustration is useful, comparing the movement of an adult animal to the growth of a plant shoot produced by its meristematic activity. Differences and similarities are seen, and I start with the latter:

> Energy is required in both. It comes from digestion of food by the animal and from photosynthesis by the plant. In both, the energy resource (Figure 35, rs) is upstream. Capturing energy at the positions 1 and I, respectively, allows later movement, later growth, downstream.

> The meristem grows and the animal moves toward more favorable positions (indicated as downstream in the figure). Immobility would result in a disadvantage to each of the competitors except in particular situations such as dormancy or hibernation. The meristem, just like the animal, explores space.

> Resources are carried downstream in the animal and in the plant. This is obvious in the animal but just as real if more discreet in the plant (Figure 36). All the material of a leafy branch flows downstream, directed toward the meristem, which organizes and produces the branch.

> Between positions 1 and 5, the animal is no longer exactly the same; it has aged. In the same way, the meristem has changed between positions I and V. In both, senescence is the term used to describe the changes.

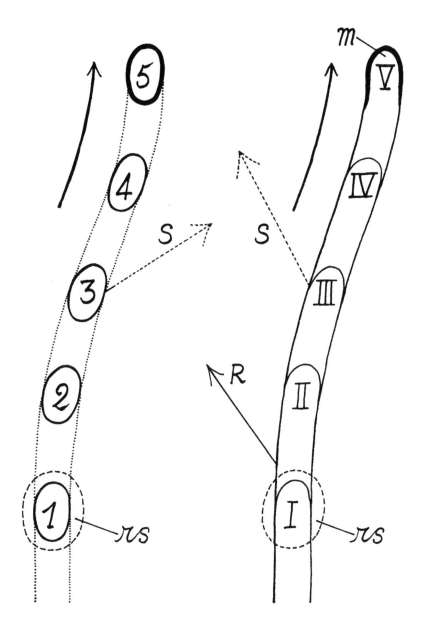

Figure 35. Movement and growth. Left, movement of an animal. Right, growth of a plant. In both cases, utilization of resources at a point (rs) allows advancement downstream, in the direction of the darker arrows. The trajectory traced by the animal is fleeting; that of the meristem (m), which constructs the plant, is long-lasting. Sexuality (S) is present in both cases; branching (R) is peculiar to the plant.

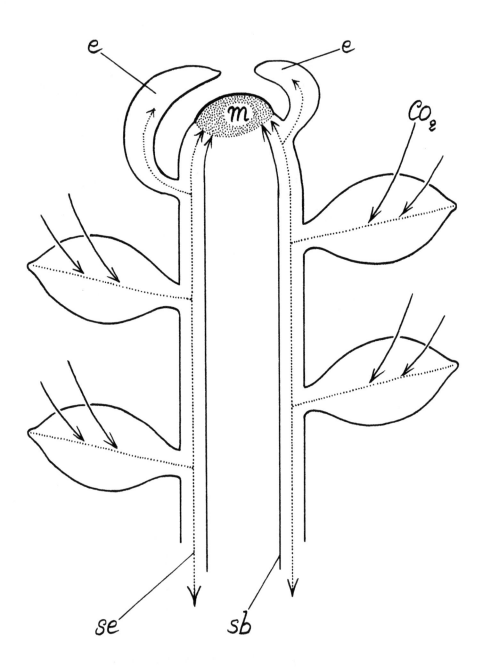

Figure 36. Movement of material toward the meristem. The meristem (m) and the leaf primordia (e) surrounding it receive water and dissolved minerals in the xylem (sb) while sugars move in the phloem (se) with translocation directed toward areas where energy is needed.

GIVEN THESE SIMILARITIES, movement and growth are different in the following ways:

> Growth in the plant is often accompanied by elaboration in the form of leaves or lateral branches (Figure 35, R); there is no equivalent of this in animal movement. Sexuality (S), the production of new individuals, present in both, is not a substitute.

> Reversing direction is possible in animals; they are able to retrace their steps. Growth of the meristem does not allow reversibility; plants are not capable of ungrowing. The size of plants can decrease through dehydration or necrosis, but these are not the same as reversing.

> The animal marks its route only episodically, leaving traces, odors, scraps, and excrement; its itinerary is fleeting. On the contrary, the meristem, producing and organizing the material of the plant, traces its entire route if it is a perennial.

> Because growth in the plant consists of the production of living tissues, it is much slower than movement in the animal. Besides, we understand that the animal's route is usually longer than the meristem's trajectory of growth, and that the plant is the prey of the animal much more often than the reverse.

The differences are greater than the similarities, but the latter are real. We can say that growth is what plants do that is most similar to animal movement, since movement is not possible for plants.

What Do the Poets Think?

Paul Valéry and Francis Ponge have some useful points for us to consider on these matters, if only to reduce the technicality of the previous discussion.

Paul Valéry (1905–07), writing about plants: "What else than the living organism allows one to see and sense true time? For a plant, a form is equivalent to an age—form is linked to size. Time is inextricably and correlatively tied to its life. A year is a node, a layer, a body separated from its surroundings and annexed, added onto, raised, directed, appointed, placed, built upon. . . ."

Francis Ponge (1942), writing about plants: "Animals move while plants unfold before our eyes. . . . Plants are an analysis in process, an original dialectic in space, progressing by the division of the preceding act. Animal expression is oral, or mimed by gestures that erase one another. The plant expression is written, once and for all. . . . Each of their gestures not only leaves a trace, just as humans do in writing, it leaves a presence, an irreversible birth *inseparable from themselves*" (Ponge's italics).

Thus, should animals be capable of movement, and not plants? The reality is subtle. An animal escapes from a zoo but it is usually captured or even killed. If an exotic plant is established in a botanical garden, it may tranquilly colonize the entire area, and the local flora may be enriched (or threatened) by the charming émigré. "Immobile from the nature of each of its individuals, the plant expands out, moving from place to place. It is by the number of its seeds, extravagantly thrown to the winds, that it expands like a fire that consumes all that it can find to devour." Valéry understood perfectly that plants may be more mobile than animals, from one generation to another.

Plants know how to use animal mobility. These plants are both useful—rice *(Oryza sativa)*, cassava *(Manihot esculenta)*, cabbage *(Brassica oleracea)*—and nuisances—ylang-ylang *(Cananga odorata)*, water hyacinth *(Eichhornia crassipes)*, witchweed *(Striga)*—unceasingly migrating with humans. As Michel Valantin (1996) humorously wrote, plants travel, courtesy of their most noble conquest, humans.

Individuals or Colonies?

A plant, *an* animal—very well! However, what does *one* signify? Is individuality the same in the two kingdoms? Plants and animals do not have the same kind of unity; to see this requires an analysis of their architecture.

First, a little about the architecture of plants: They elaborate themselves according to strict rules, rules that lead toward the conceptualization of models. The dynamics of these models is the subject of the study of plant architecture (Hallé and Oldeman 1970, Hallé et al. 1978). Only in the humid Tropics, where gravity is the only significant physical constraint, is it possible to see the full range of architecture known to occur in vascular plants (see Figure 79, Chapter 6). Is there an animal architecture comparable to that of plants? The general organizational plans separating the zoological phyla are well known. Are they comparable to the architecture of plants in some manner? The answer is no, which is understandable because of the different means of capturing energy: an external surface for assimilating solar energy in plants, an internal digestive surface in animals.

The architectures of the two kingdoms are not comparable, as if they were situated at different levels, but that does not hinder our understanding of the architecture of units versus colonies. In zoology, the distinction between single units and colonies is classical knowledge, known to Aristotle in the 4th century B.C. The concept of colonial animals was known to the Swiss naturalist Abraham Trembley at the beginning of the 18th century; he observed hydras budding in fresh water. The concept was assumed by the French zoologist Édmond Perrier, who attempted an early classification of colonial animals at the end of the 19th century. In zoology, the contrast between the unit and the colony has always been clearly understood. Freely moving animals—locust, ray, gecko, human—are all units. Colonies exist only in some aquatic animal groups, sessile for the most part: hydras, corals, some medusas, bryozoans, tunicates, ascidians (Boardman et al. 1973). Established through the loss of individuality, the colonial mode of life is considered as providing a competitive advantage for the species that have adopted it.

In plants, the situation is much simpler, and a reference to epistemology makes it easier to comprehend. Only in the 19th century did some researchers understand that plants might also be of a colonial nature. After disembarking from the *Beagle*, Charles Darwin wrote (in Fitzroy 1839) that it seemed astonishing that distinct individuals could be united, yet each tree confirms the fact, that in effect its buds must be considered individual plants. The polyps of a coral or the buds of a tree are examples in which separation between individuals has not been completed. Jean-Henri Fabre (1823–1915), whose timeless studies in entomology are remembered by all naturalists, also clearly perceived the colonial nature of plants: "It is said that a plant is comparable to a coral head covered with polyps; it is not a single being but a collective one, an association of individuals, all relatives, all closely united, involving one another and working toward the prosperity of the ensemble. It is, same as a coral, a sort of living hive whose inhabitants have a life in common" (Fabre 1996).

Just what is this elementary individual, constituting the "collective being" that is a plant? Fabre's reasoning was admirable. First, he remembered that an individual defines itself as "all that forms a living unity and cannot be divided without losing life." An examination of horticultural practice—cuttings, grafting—gave him the solution: "A tree can be subdivided into distinct new plants that produce branches; in turn, the branches can continue to grow as long as they contain buds, but the bud itself is not divisible; it is destroyed by being split up. The vegetable individual is thus a bud" (Fabre 1996).

The Discovery of Reiteration

The vision of the colonial nature of plants disappeared for nearly a century before resurfacing thanks to research in tropical botany. In 1972, while paddling a pirogue on the Yaroupi River in French Guiana, the Dutch forest botanist Roelof Oldeman became aware of the similarity between a tree and the branch he was holding. Certainly, branches, or suckers, known by arborists for a long time, were so familiar that nobody paid them any attention. A branch required the perspective of a researcher, perhaps also many hours in a pirogue on the smooth water between two forested walls, to see the reality: It is a young tree growing on an old tree (Oldeman 1972). Oldeman then demonstrated that most trees thus have the possibility of accumulating architectural units that grow on others, forming a colony (Figure 37). A true story: When he was preparing his thesis on the growth of trees in rain forests of French Guiana, Oldeman was so charged with this idea that one night he was awakened by a telling nightmare—his fingers and feet were in the process of forming colonies (Figure 38). Oldeman (1974) gave the name reiteration to this process of becoming a colony. In trees, even those of temperate climates, reiteration is often easy to see.

Without going into the details of the phenomenon of reiteration, already well studied, the following essentials will suffice. At the beginning of its life, a tree has a single architecture. Later, and earlier when light is abundant, other architectural units add to the first and the tree continues to accumulate these reiterated units, or reiterations. As constituent elements of the colony-tree, with time these reiterations become more and more numerous, and smaller and smaller, while their form is simplified to the point that they end up reduced to a leafy stem with flowers. This is how trees grow, redwoods *(Sequoia, Sequoiadendron)* and lignum vitae *(Guaiacum)*, peach

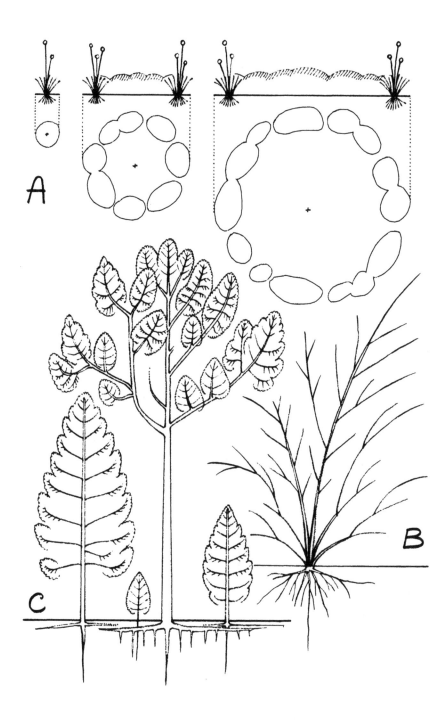

Figure 37. Plants with colonial architecture. (A) A clone of grasses. (B) A shrub, for example, *Buddleja*. (C) A large tropical Asian tree, *Dipterocarpus* (Oldeman 1974, Edelin 1990).

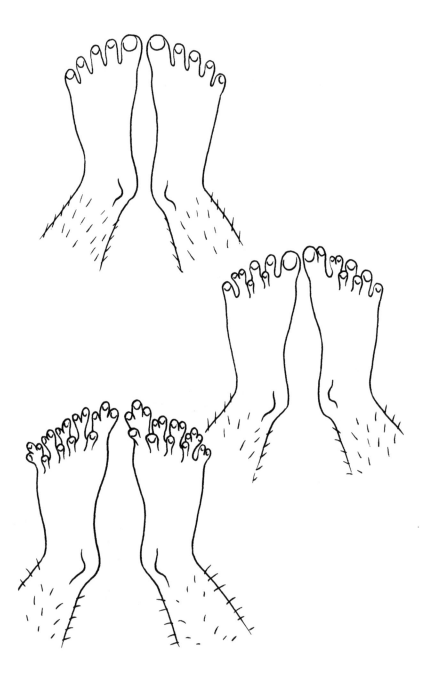

Figure 38. Oldeman's nightmare. The author of the concept of reiteration described the dream he had as he completed his doctoral dissertation in Montpellier in 1972 (Oldeman 1974, with his friendly permission).

tree *(Prunus persica)* and parasol pine *(Sciadopitys verticillata)*, olive *(Olea europaea)* and plane *(Platanus)*.

The moment has come to ask some difficult questions. Are trees individuals? Is an individual, instead, the reiteration? And first, just what is an individual? If the idea makes sense, is an individual the same for both plants and animals? The great majority of scientists evades this genre of question, which touches on cultural fundamentals, qualifying such inquiries as "philosophical." This word, which they use with condescension, has always inspired me. Contemporary scientific practice is encumbered by problems that reveal themselves as secondary. It is too bad science evades this question, which would lead us to the roots of individual existence. I cannot pretend to be a philosopher, but such deep significance interests me. Albert Szent-György wrote that when he went fishing, he always used a big hook, because although he knew he would catch nothing, he found it much more interesting not to catch the big fish than not to catch only the little ones.

What Is an Individual?

What provokes me to think this is not a useless question is the use of *individual* to describe a plant, an animal, and a bacterium equally. It was liberally used by Darwin and Fabre during the 19th century. It is still in use in a diversity of contemporary research areas: forest ecology, bacteriology, tree architecture, immunology, population genetics. Is it legitimate? I do not pretend to be in a position to define the individual, but at least I would like to reassemble some criteria traditionally used by biologists in making such a definition.

The first criterion is etymological: An individual is a living being who cannot be divided into two equal parts without dying (Errera 1910, Souèges 1935). The second is genetic: An individual has a genome that is stable in space and time, in space because the genome is the same in all parts of the organism, and in time because the genome is the same throughout the life of the organism (Slatkin 1987). Finally, there is an immunologic criterion: An individual is a functional entity, clearly delimited, singular, different from others, and possessing a recognition of self and nonself (J. Bernard et al. 1990). Animals, since they fulfill these requirements, are individuals. However, does the application of this term to plants reveal a zoocentrism? Fabre (1996) said, "Relative to animals, in the immense majority of cases, to divide is to kill; relative to a plant, to divide is to multiply."

If we divide a plant into two parts, treating the two halves by standard techniques, two new plants are produced. Most plants are susceptible to manipulation by well-known horticultural techniques: dividing, suckering, cutting, layering, grafting, etc. For certain plants, such as lilac *(Syringa)* or manioc *(Manihot esculenta)*, such multiplication is common. For others, such as banana *(Musa)*, ginger *(Zingiber officinale)*, or common bamboo *(Bambusa vulgaris)*, it is the only known means of propagation. Finally, several extraordinary plants, such as the killer alga *(Caulerpa taxifolia)* or water hyacinth *(Eichhornia crassipes)*, can be ground up without causing their death, each piece being able to reproduce a new plant. It is clear that plants are not individuals in the etymological sense. If a human could be regenerated by such means (Figure 39), which would we consider the individual?

The genetic criterion of individuality remains little explored. Old discoveries reconfirmed by more recent results indicate that the genomes of many plants are unstable in space and time (Chapter 5). So, plants also may not be considered individuals in the genetic sense.

The third, immunologic criterion makes us recall some interesting results. The differentiation of self and nonself is the central problem in this discipline (Burnet 1969). From arthropods to vertebrates, animals have an immune system that protects them from disease. In humans, all our defenses against nonself involve a system of tissue types, equivalent to blood types, that determine the success of organ transplants or grafts: "Our tissue type is our biological identity card, the seal of our individuality and . . . the guarantee of our integrity" (Dausset 1990). With immune mechanisms relying on the mobility of the cells charged with surveillance—lymphocytes, for example—it is *a priori* improbable that plants would incorporate such a mechanism. Their cells being immobile, plants do not have an immune system (Dausset 1990).

However, plants do know how to recognize self and nonself. Wheat pollen falling on the stigma of a zucchini flower will be recognized as nonself and will not germinate. In other plants, such as the sweet potato *(Ipomoea batatas)*, pollen from a flower falling on the

Figure 39. The clonable man, in the style of Jorodowsky and Moebius (1981).

stigma of that same flower will be recognized as self and not germinate, whereas pollen from another plant of the same species will be allowed to germinate. This is called self-incompatibility. Plants maintain a way out: In a normally self-incompatible plant such as the bee orchid *(Ophrys)*, if the pollinating insect has not been effective and the pollen of another plant has not been deposited, as a last resort the plant will accept its own pollen. Thus the limits between self and nonself may not be so strict.

In fruit tree culture, grafts between different varieties of a species are frequently accomplished even between quite different varieties. Grafting between different species is much more difficult but possible by highly skilled horticulturists. Manual skill is not enough; sensitivity to environmental conditions and an acute awareness of plant health are also important for this sleight of hand. An exceptional grafter can even accomplish fusions between different genera of the same family (Chaudière 1992), such as an olive *(Olea)* on an ash *(Fraxinus)*, or a chestnut *(Castanea)* on an oak *(Quercus)*. I even know of a successful graft between two distinct families: a member of the Didiereaceae from southern Madagascar grafted on a member of the American family Cactaceae, mirroring the evolutionary proximity of the two families (René Hebding, personal communication).

We see that the difference between self and nonself is not so strict in plants. Does the immunologic criterion suggest that plants are individuals, or do the two other criteria (vegetative division and genomic stability) suggest the opposite? Before laying out a balance sheet, let us try to look clearly into the question that led to this discussion of the notion of the individual.

Is a Tree an Individual?

We have seen that a tree can be split apart without dying. From this point of view, we must answer no to the question, Is a tree an individual? Moreover, there is a difference between the reiterations that occur naturally in a tree crown and those of the horticulturist who propagates the tree vegetatively. The ensemble of reiterations in a tree forms an integrated entity, borne on a single trunk. Is it conceivable that a particularly well integrated colony could emerge as an individual? Would the individual thus be a recursive idea, a nested reality? Let us see what the zoologists think. They have a much longer familiarity with colonial structures than botanists, who are neophytes in these matters.

In colonial animals, whose individual constituents are zooids, two complementary tendencies have been identified: decrease in size of the zooid with increase in size of the colony (Beklemishev 1970). In sponges, two levels of individuality are recognizable, at the level of the cell and also that of the entire structure (Rasmont 1979). During the evolution of colonial animals, the process of integration among zooids was associated with the emergence of perfectly organized colonies as individuals with superior integration (Coates and Jackson 1985). It is necessary, as Rasmont (1979) has observed, that we "renounce dogmatic dichotomies." We can allow a tree with reiterations to be an individual and a colony at the same time. However, a detail enters here. The individuality we are contemplating is that of plants, little integrated, divisible without damaging the whole, and

thus very different from what zoologists discuss. A tree's performance is not comparable to an animal's but is closer to that of a population of animals.

We revisit reiteration. It is not limited just to trees (Figure 37). It
is also applicable to shrubs (hazelnut or lilac) and most lianas (bou-
gainvillea or clematis) as well as many herbs (carnations, *Gaura, Rus-
selia*, banana, iris, fescue). Reiteration may be confused with vegetative propagation in plants. Each reiteration establishes its own root
system and thus becomes capable of leading an independent life.
There is no further difference between colonial architecture and a
clone. This is equally true for certain trees, such as olive or aspen, that
are capable of forming clones that grow very old, thus destroying the
distinction between reiteration and vegetative multiplication.

With colonial architecture being identifiable this way, it becomes
possible to define unitary architecture *a contrario*. In temperate regions, unitary architecture is found in annual plants—scarlet pim-
pernel *(Anagallis arvensis)*, wall cress *(Arabis)*, puncture vine *(Tribu-
lus terrestris)*—or biennials—mullein *(Verbascum thapsus)* or honesty
(Lunaria annua). These plants have by definition a single architectural unit. Certain tropical plants are organized into a giant architectural unity. This is true of *Agave* (whose stems flower only once),
Cecropia, and columnar palms (Figure 40).

One final time: Plants and animals function in contrasting ways.
Plants are primarily colonial, with unitary ones probably not more
than 20% of all the species. The situation is the opposite in animals,
which are mostly unitary, with colonial architecture mostly limited
to a number of aquatic or marine species that are usually fixed in
place. It is disturbing to extend colonialism to free-living animals
graphically; if the result provides some aesthetic interest (Figure
41), it does not give the impression of being viable.

Potentially Immortal Beings

The importance of distinguishing the unit from the colony also
raises questions of longevity. In plants as in animals, unitary organisms have lifetimes roughly proportional to the size of the body,
generally not so long. Many plants and animals live only a few years,
a few months, even just a few days. The maximum would be three
centuries, a duration attained by some unitary trees of large size,

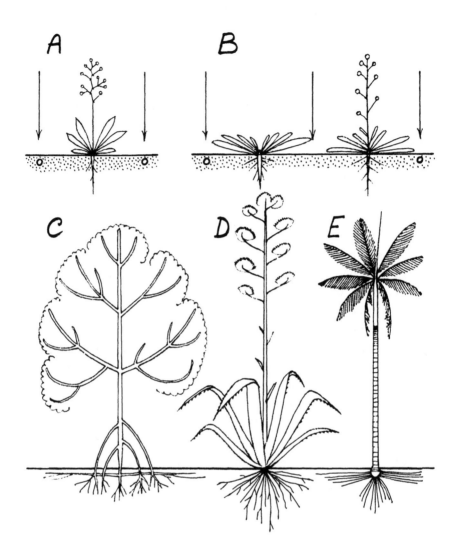

Figure 40. Plants with unitary architecture. (A) Annual plant. (B) Biennial plant. Vertical arrows indicate winters, passed in the form of a seed, sometimes a rosette. Plants with unitary architecture can sometimes attain great size: *Cecropia* (C), *Agave* (D), columnar palms (E). Compare with Figure 37, plants with colonial architecture.

Figure 41. Colonial humans, from a tapestry by the Ibans of Borneo (Sellato 1989).

such as Australian araucarias, or some slow and large animals, such as the giant tortoises of the Galápagos Islands.

Colonial architecture may confer an indeterminate life span on organisms; using a misleading term, it could be said that colonialism makes them potentially immortal. That is to say that their length of life is only limited by unfavorable external conditions: earthquakes, bad weather, diseases, large predators. The age of some corals has been estimated as 15,000 years, and similar ages have been verified for colonial plants. Pliny the Elder (A.D. 23–79) marveled at the "enormity of the oaks of the Hercynian forests, contemporary with the origin of the planet" and in a "nearly immortal condition" (Lieu-taghi 1991). The Swiss botanist Alphonse de Candolle affirmed the idea of the immortality of plants in the mid-19th century. It is not a shocking idea, and Fabre (1996) the entomologist expressed it with his usual insight: "If it is really a collective being whose successive generations overlay one another, a tree must live for a very long time and not perish except from accident, since old buds succeed each other year after year, maintaining a vegetable community always young and ready for the future."

Watkinson and White (1986) have noticed that many works on senescence pass over plants very superficially or only mention annuals. It is, they wrote, "modular" plants (what I here call "colonial") that do not become senescent since they are potentially immortal. The facts speak for themselves, and the life span of certain plants is prodigious. I borrow facts that have the shortcoming of concerning only exceptional "individuals" but that have the merit of being reasonably trustworthy (Table 1).

The data in Table 1 call for two comments. First, those interested in the ages of plants choose to emphasize the study of trees, censused as living much longer than herbs. That is debatable. Although much more discreet, herbs should have much greater hope for long life. Vegetative propagation would give them a propitious mobility to exploit new habitats. Second, if in nature virtually no plant is immortal, human activity could remedy that. Consider the banana (Musa). The commercial varieties do not produce seeds, and their culture is limited to isolating sprouts and placing them in favorable ecological conditions. These banana clones are as ancient as tropical agriculture itself, 10,000–20,000 years, and their death would be a great loss; sometimes they are intentionally planted in

Table 1. Potentially immortal beings

1,000 years	baobab *(Adansonia digitata)*, jujube *(Zizyphus jujuba)*, litchi *(Litchi chinensis)*, clones of vegetable sheep *(Psychrophytron [Raoulia] mammillis)* or eagle fern *(Pteridium aquilinum)*
2,000 years	redwoods *(Sequoia, Sequoiadendron)*, New Zealand kauri *(Agathis australis)*, dragon tree *(Dracaena draco)*, Australian cycads, yew *(Taxus)*, olive *(Olea europaea)* of Roquebrune-Cap-Martin in the Maritime Alps, chestnut *(Castanea sativa)* of 100 Horses on Mount Etna
3,000 years	*Ginkgo biloba*, cedar of Lebanon *(Cedrus libani)*
4,000 years	*Taxodium mucronatum* at Santa Maria del Tule, near Oaxaca, Mexico, *Fitzroya* in Chile
5,000 (–10,000) years	bristlecone pine *(Pinus longaeva)* in the White Mountains of California
10,000 years	clones of aspen *(Populus tremuloides)* in Utah
11,000 years	clones of creosote bush *(Larrea divaricata)* in the Mojave Desert of California
13,000 years	clones of huckleberry *(Vaccinium)* in Pennsylvania, alive at the end of the Pleistocene and contemporaries of the last Neanderthals

Data from Heywood (1985), Buss (1987), Bourdu and Viard (1988)

ecologically unfavorable sites although the global distribution of commercial varieties reduces such risks. For all cultivated bananas to die would require that humans stop eating bananas. That is highly improbable. To me, the banana, though not a tree, represents an optimal means of immortality for colonial or clonal plants. It is not the only example. Manioc *(Manihot esculenta)* and potatoes *(Solanum tuberosum)*, cultivated clonally year after year, are also immortal for the same reason as the banana.

Let us dwell on this term, immortality. For both plants and animals it is the colony itself, not its constituent elements, that has the capacity for immortality. A reiteration in a tree crown, a banana sprout, a sucker from a cereal, a coral polyp—these do not live as long as a unitary being of comparable size. We find comparable situations in the biology of social insects (Chapter 6): An ant lives for a short time but ant colonies are potentially immortal (Holldobler and Wilson 1996). Fabre (1996) wrote, "The individual perishes but the society persists." Plants and animals are thus confronted by the choice of remaining an individual or becoming a colony, with enormous consequences for their longevity. For animals, this choice has an additional consequence: Only the unit is mobile and the price paid for mobility is a short life. This leads us naturally to speak of death, and an unexpected implication.

Two Ways of Dying

Plants and animals do not die the same way. An animal is alive or dead. It may be difficult to determine if it is one or the other, but we cannot imagine its being both at the same time. During the course of its life, most cells die and are replaced by new cells, but the turnover is at the cellular level. Even if it molts or goes through a metamorphosis, this does not alter the fact that the animal is either alive or dead.

Plants do not have such cellular turnover. Their cells are fixed in place and thus cannot be replaced. If we make an exception for specialized annual and biennial species, plants know (in contrast to animals) how to get rid of organs that no longer serve them. Falling leaves and the dropping of low branches do not have true equivalents among animals. Sloughing off the surface of the skin in snakes, seasonal changing of feathers in birds, and losing spines in porcupines are phenomena of a scope more limited than self-pruning, resulting from the specialized characteristics of the tissues disposed of by these animals. In plants, it is as if continual death of used organs has replaced death of the plant itself, the plant having acquired immortality through colonialism. A bad side effect must be avoided. Dead leaves and fallen branches are never actually replaced because their points of attachment remain vacant. Indeterminate morphogenesis allows the plant to elaborate other leaves, other branches.

Contrary to animals, most plants are capable of dying at an extremity while continuing to grow at another point. Clonal plants—eagle fern (*Pteridium aquilinum*), date palm (*Phoenix dactylifera*), lily of the valley (*Convallaria majalis*), buttercup (*Ranunculus*)—have the

Figure 42. Two ways of dying. The possibility that colonial plants live and die at the same time is illustrated by a cork oak, shown here after damage from fire (Roger Prodon, personal communication). Transferred to an animal, such behavior is clearly grotesque.

same strategy as colonial trees—oak *(Quercus)*, meranti *(Shorea)*, manil *(Symphonia globulifera)*, tembusu *(Fagraea)*, *Eucalyptus*. Their structures weakly integrated, their reiterations relatively independent from each other, these plants have the capacity simultaneously to die and to live without our noticing a behavior so different from ours. Transfer the strategy of an evergreen oak to a brave dog; the grotesque result shows how different are the manners of death (Figure 42).

WE RETURN TO LIVING. A vast domain of function reveals numerous and fundamental differences between plants and animals (Chapter 4). But first it is necessary to make a detour into the realm of the cell. Animal and plant cells reveal their differences, indispensable to understanding how living beings work.

CHAPTER 3

The Cell

I will confine myself to the world of plants, because their cells, provided with rigid walls, have the look of dressed stones placed by a mason-architect.

In the course of construction of animals, their flabby cells swarm and migrate while forming layers that fold on themselves into grooves, tubes, vesicles, nodules. These form organs in a precise fashion, but not with the classical simplicity of the embryogenesis of plants.

MICHEL FAVRE-DUCHARTRE, *Unions Créatrices*, 1997

Because of my insensitivity and reduced nervous system, humans claim that I am, despite my grand appearance, a collection of cells rather than a true individual. Thanks for the compliment. If to become a being in their eyes is to pay the price of having a brain, nerves, and blood, and to look after them, I don't want these, and never did.

MICHEL LUNEAU, *Paroles d'Arbre*, 1994

Telegraph Line
You see this telegraph line at the back of the valley, whose
    rectilinear trace cuts the forest on the facing
    mountainside.
All the posts are iron.
When first put in, the posts were wood.
After three months they began to grow branches.
They were pulled up, returned to the soil and planted upside
    down, roots exposed to the air.
After three months they again produced new branches,
    established roots, and continued their lives;
It was necessary to pull them all up and establish a new line
    by ordering, at exorbitant expense, those iron posts from
    Pittsburgh.

BLAISE CENDRARS, *Au Coeur du Monde*, 1924

T HE WORD CELL EVOKES A PRISON. To study cells is austere
enough because the scale at which they are observed denies us
any practical, direct experience of them. The cell is the key level,
however, upon which all else rests. The cellular domain is also full of
surprises because cells foreshadow the organisms that they constitute. Cells have been so well described by specialists (Alberts et al.
1983) and for the general public (Rosnay 1988) that I will not go
into the details of cellular ultrastructure or organelle function. I
accentuate the differences between cells of plants and animals, but
since the differences will become clear, I start with the similarities.

Characteristics of the Eukaryotic Cell

It is a fact that both plants and animals are composed of cells with
nuclei and for this reason are known as eukaryotic organisms. Fern,
slug, lilac, flying squirrel, human being—members of the two kingdoms are vast aggregates of cells, each of which contains a nucleus.
It is certainly not by chance that the eukaryotic cell has allowed the
emergence of all forms of complex life. Cells without nuclei, pro-
karyotic cells—those of bacteria, for example—seem not to have
been capable of permitting such complexity. They have played a
useful and even decisive role, however, as we shall see.

Did the eukaryotic cell appear only once in the course of evolution? If all eukaryotic cells had a common origin, that would explain
the many similarities uniting animal and plant cells. Animal and
plant cells have a goodly number of structures and organelles in
common (Figure 43). These are neither plant nor animal but can be
regarded as part of the fundamental organization of the eukaryotic
cell. I cite about ten of the more important structures, briefly
describing their functions.

The most remarkable organelle, at the heart of the nature of the
eukaryotic cell, is the nucleus: unique, spherical, 5–10 μm in diameter, defined by a double membrane pierced by pores. It contains
chromosomes composed of the famous deoxyribonucleic acid (DNA)
and associated proteins. It also contains the nucleolus, a spherical
mass of granules of ribonucleic acid (RNA) and partially assembled

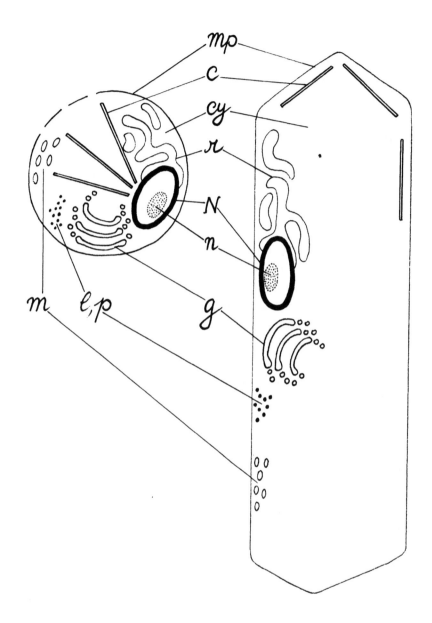

Figure 43. Characteristics of eukaryotic cells. Left, an animal cell, 10 μm in diameter. Right, a plant cell, 100 μm long. Structures and organelles common to both cells: mp, plasma membrane; c, filaments and tubes of the cytoskeleton; cy, cytoplasm; r, endoplasmic reticulum; N, nucleus, containing the chromosomes; n, nucleolus, with ribosomes; g, Golgi apparatus; l, lysosomes; p, peroxisomes; m, mitochondria.

ribosomes. The central organizer of the cell, the nucleus, contains the plan necessary for the synthesis of all the chemical constituents of the organism, plant or animal, in the form of the two nucleic acids (Chapters 4 and 5). Associated with the nucleus, the endoplasmic reticulum is a vast network of plates, sacs, and membranous tunnels, playing a central role in the synthesis of needed proteins and lipids.

The Golgi apparatus is a stack of flat sacs, producing numerous small, mobile vesicles. It assures the sorting out, packaging, and transporting by the vesicles, and thus the delivery, of macromolecules destined for all the structures and functions of the cell. The mitochondria, small and numerous, have the dimensions and appearance of bacteria. They assure respiration for the entire organism; they burn glucose in the presence of oxygen, liberating water, carbon dioxide, and a significant amount of energy. The latter is in the form of adenosine triphosphate (ATP), responsible for most of the functions of the organism. I return to the mitochondria in a discussion of biochemical processes in plants (Chapter 4).

In eukaryotic cells, mitochondria have a special significance: They truly are bacteria! Primitive, anaerobic bacteria are incapable of respiration in the presence of oxygen, but the integration of aerobic bacteria with such capability into the cell permitted the emergence of cells fit for survival in a world that had become rich in oxygen. Rather than digesting these aerobic bacteria, the cells nourished and maintained them symbiotically, using their capability of consuming oxygen and producing energy, just as we raise cows, who turn hay into milk (Alberts et al. 1983). That also explains why mitochondria contain their own DNA and have partial autonomy in controlling their own division.

Lysosomes are minuscule vesicles (0.5 µm in diameter) containing enzymes involved in cellular digestion. In plants, the function of lysis is provided by young vacuoles, phytolysosomes. The products of digestion escape by crossing the lysosomal membrane and are reused by the cell. Peroxisomes have the function of detoxification; they are vesicles with the same dimensions as lysosomes and contain oxidative enzymes that break down hydrogen peroxide and free radicals.

The cytoskeleton is a network of filaments and protein tubules, giving form to the cell, permitting it to move its organelles around, and making cells mobile. All the organelles are immersed in a gelatinous and transparent fluid, the cytoplasm, which is surrounded by a membrane that assures the individuality of each cell and separates it from the exterior world.

These characteristics of the eukaryotic cell are extraordinarily ancient; they certainly existed prior to the divergence of plants and animals some 700 million years ago (Margulis and Schwartz 1988). The organelles have been known for the most part for only a century or so. Aristotle was not acquainted with cells. Antoine-Laurent Lavoisier never heard any discussion of the nucleus. Claude Bernard and Louis Pasteur were ignorant of mitochondria and DNA even if they had intuitions of their existence.

An odd similarity between plants and animals appears at the cellular level. We know that organisms differ prodigiously. Length varies by a factor of 10^8, mass by a factor of 10^{24}, from smallest to largest. Vogel (1988), from whom I have borrowed these statistics, noticed that whatever the conditions in which they are found and in whatever organism, cellular dimensions do not vary much. They are of the order of 10 μm in a flea or a whale, and about 100 μm in a baobob tree or a duckweed.

Plants and animals are not the only living creatures composed of eukaryotic cells. They also constitute fungi, protists (single-celled organisms), but not bacteria, as mentioned. As for viruses, they are not even cells; viruses are not considered to be complete living beings but rather nucleic acids that have escaped their original cells and become enclosed in a protein capsule (Margulis and Schwartz 1988). Alternatively, viruses can be considered along with prions as beings that have invented an original solution to the problem confronted by all that live: reproduction (François Bonhomme, personal communication).

Differences in Structure

The cellular level, appropriately considered by biologists as the one least marked by the evolutionary divergence that separates plants from animals, does reveal a number of significant differences between the two kingdoms. That both plants and animals are composed of eukaryotic cells does not mean that their cells are identical. They are the same at the level of fundamental molecular mechanisms of which I say nothing here: DNA replication, protein synthesis, production of mitochondrial ATP, the Krebs cycle, membrane structure, etc. At a higher level of integration, however, that of structure and function, remarkable differences appear between plant and animal cells (Figure 44). For a biologist, even a beginner, it is scarcely possible to confuse the two kinds of cells.

A first difference is a consequence of the presence or absence of a wall. In animals, the cell is destitute of a wall; the cell is flabby and naked, delimited only by its membrane. In plants, the cell is surrounded by a wall, which produces the form of a rigid box made from a typical vegetable material: cellulose. Cellulose is, as a result, the most prevalent organic molecule on the surface of the Earth. Cellulose accounts for more than half of all terrestrial biomass (Robert and Roland 1989). Even the book you hold is made from fibers of cellulose.

In animals, the cell is small, naked, and potentially mobile, at least during the embryonic phase. In plants, the cell is much larger and immobile because its biology is dominated by the cellulosic wall. There are some exceptions to these two extremes. Certain plants—algae, mosses, ferns, ginkgo—produce swimming gametes that are naked cells. Some animal cells produce a covering, the glycocalyx, outside their plasma membrane that facilitates cellular recognition. We shall see other examples of such ambivalence in the two kingdoms. Even if plant and animal solutions are different, even opposed, each of the two groups is capable of producing practically all that the other can make.

Supple when the cell is young and still small, the cellulosic wall accommodates cell growth, then loses its flexibility and encloses the adult cell in a veritable external skeleton. This provides strength but also prevents migration of cells in the plant body. Plant cancers do not metastasize. On the contrary, cellular migration is a major phenomenon in animal embryogenesis.

Water containing dissolved substances attracts pure water. This is a force of extraordinary power, resulting in pure water diluting such solutions. The pressure of diffusion through a membrane, called osmosis, causes an animal cell to burst if it is placed in distilled water. In plant cells, the presence of the wall results in a very different osmotic response. Placed in distilled water, the plant cell limits the entry of water by the pressure it exerts on the wall. The cell does not burst because it is physically contained within a rigid wall (Figure 44) against which it exerts pressure. The cell becomes

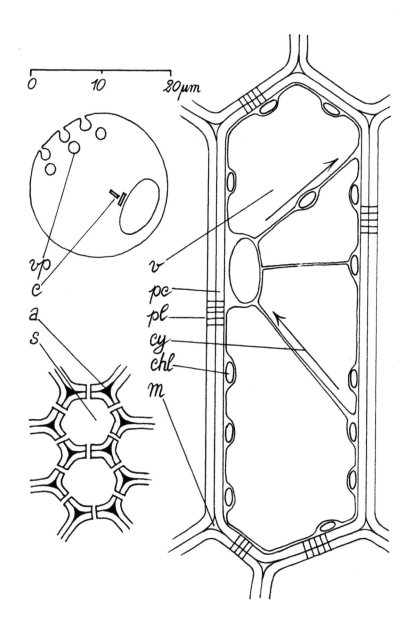

Figure 44. Differences between an animal cell (left, 10 μm in diameter) and a plant cell (right, 100 μm long). Organelles or functions unique to animal cells are phagocytic vesicles (vp) and centrioles (c). Those unique to plant cells are the vacuole (v), cellulosic wall (pc), plasmodesmata (pl), cytoplasmic current or cyclosis (cy), chloroplasts (chl), and ducts called lacunae (m). All these ducts and spaces together constitute the apoplasm (a) whereas the symplasm (s) is the cytoplasm common to all cells of the plant, connecting from one cell to another via plasmodesmata.

fully turgid at its maximum hydration (Cruiziat and Le Guyader 1990). Such turgor is vital for plants; it is the principal force responsible for the mechanical rigidity of young tissues. We are unable to hold up an empty balloon, but inflated, there is no problem.

Plant cells are not as isolated from each other as the presence of a wall might make us believe. Each cell communicates with its neighbors by tunnels, or plasmodesmata (1000–10,000 per cell), that permit the passage of cytoplasm from one cell to another (Figure 44). This is more than communication. In plants, the cytoplasm as well as the plasma membrane is common to all the cells of the organism. The notion of a symplast, including the cytoplasm of the entire organism, is almost exclusively vegetable.

Despite plasmodesmata, the cell wall is an obstacle. Its porosity limits exchange between plant cells and their surroundings. Plant cells must be nourished by small molecules, and similarly, the intercellular signal molecules (auxins, cytokinins, gibberellins, ethylene) necessarily are of modest dimensions. Much larger, they would not be capable of crossing the cell wall. The wall surrounding the cell provides opportunities for the plant because its mechanical properties make possible plants of great size and, above all, with vast surfaces capable of filling space. One of the essential characteristics of plants, and a major difference from animals, is a large surface in relation to a modest volume.

The enormous vacuole, which occupies as much as 90% of the volume of the plant cell and which is a bag of liquid limited by a membrane, has no equivalent in animal cells. Vacuoles develop in young cells immediately after mitosis. They are first implicated in the digestive process, or lysis, at which stage they are called phytolysosomes. Following this, while absorbing water and producing a turgid state, vacuoles become directly responsible for cellular growth (Figure 44), which is made possible by the flexibility of the young wall.

Plant cell growth is truly spectacular. The animal cell measures about 10–20 μm in diameter and is not observable without magnification. Adult plant cells, some 100 μm long, are easily visible to the naked eye. The dimensions given here are typical; cells several millimeters long, even several centimeters, occur in plants as well as in some animals.

For nutrients, enzymes, or message molecules, diffusion is easy and quick in the small animal cell, whose cytoplasm can be crossed in a small fraction of a second. On the contrary, in the huge plant cell, diffusion through the liquid phase would be too slow to be effective. There is a cytoplasmic current, or cyclosis, a circular movement that stirs material throughout the thin cytoplasmic layer that separates the vacuole from the wall. Animal cells apparently do not have such movement.

The vacuole promotes plant cell growth at little cost of energy, with little biochemical synthesis, and with the simple requirement of water. That is not the only function of the vacuole; it also stores useful molecules or wastes. To rid itself of toxins, the plant cell mainly stores them. Little by little, vacuoles become filled with various chemical compounds: calcium oxalate, anthocyanins, flavonoids, cyanogenic compounds, etc. It has been said that plant cells live dangerously (Matile 1975), a little like a town that fills abandoned subway stations or sewers with discarded dioxin. These plant vacuoles may be thought of as "bombs" stored for use against aggressors in a "chemical war" (Jean-Claude Roland, personal communication). Besides storage, the essential role of the vacuole is cell growth, and growth of the entire plant.

One Cell Within Another

Plastids are cell organelles limited to plants and a few protists: *Euglena*, peridinians, diatoms, and *Volvox*. Fungal cells are devoid of them, as are animal cells. The presence of plastids is, along with the vacuole and wall, one of the principal characteristics differentiating plant from animal cells. As strange as the idea is, we must consider plastids as small cells inside the cell. They lack a true nucleus, sometimes contain chlorophyll, and in the latter case, resemble cyanobacteria integrated into the plant cell.

Integration of a small bacterial cell into a large cell has already been used to explain the presence of mitochondria. Animal cells remained at such a stage. In plant cells, in addition to mitochondria, cyanobacteria were transformed into plastids in a separate ingestion. Like mitochondria, they have their own DNA and partial control over their own division. This symbiosis, between cells encased within one another, has been recognized as a general phenomenon by Margulis and Fester (1991), who revealed its importance. Without bacteria there would be no mitochondria or plastids, thus no eukaryotic cells. Plants and animals would not exist since they owe their existence to these symbiotic bacteria. This leads to a revolutionary idea defended by Stephen Jay Gould (1996): Bacteria have been, are, and will remain the dominant group of organisms on our planet. Essentially subterranean, living off the chemical energy contained even in rocks, they are also sustained in the atmospheric milieu through the subterfuge of intracellular symbiosis in two unimportant superstructures: plants and animals. In this vision of evolution, even human beings are a sort of efflorescence of bacterial origin!

Let us return to plastids. We recognize three kinds. The most important are chloroplasts, full of chlorophyll and thus responsible for the green color of plants. Using chlorophyll, carbon dioxide, water, and solar energy, chloroplasts synthesize sugars, in a sense the exact opposite of mitochondrial respiration. To do this they consume carbon dioxide, transform solar energy into the chemical energy of glucose, and release the oxygen that allows us to respire. Plant cells also contain amyloplasts, which store starch, and chromoplasts, charged with pigments that provide the vivid colors of flowers and fruits. Thereby, animals are attracted to pollinate flowers, and consume fruits and disperse seeds (Chapter 4).

Thus plant and animal cells have many characteristics in common yet some differences of major consequence for the biology of each kingdom (Table 2). The cell wall, symplast, vacuole, and plastids are the four principal structures distinguishing plant from animal cells. Could it also be overall complexity? I rally to the idea of Trewavas (1986) for whom the plant cell is "probably more complex" than the animal cell. In fact, the plant cell carries on nearly every function of the animal cell, adding to those the synthesis of chlorophyll, photosynthesis, and the production of secondary compounds (flavonoids, phytochrome, carotenoids, coumarins, lignin, anthocyanins, alkaloids, etc.), which are much more numerous and diversified than the secondary metabolites of animal cells (Chapter 4).

Differences in Function

The differences between plant and animal cells are not merely structural but functional. There are five differences that appear at a glance; they concern cellular mobility, assimilation of food, divi-

Table 2. Structures that plant and animal cells have in common, and
structures that distinguish them

Plants	In Common	Animals
cell wall	plasma membrane	glycocalyx
symplast	cytoplasm	
	nucleus	
	chromosomes	
	nucleolus	
	endoplasmic reticulum	
	Golgi apparatus	
	mitochondria	
	lysosomes	
	peroxisomes	
	ribosomes	
	cytoskeleton	
vacuole		
plastids:		kinetic apparatus:
chloroplasts		centrioles
amyloplasts		asters
chromoplasts		

sion, differentiation, and totipotency. Cellular mobility has been
discussed as a difference in behavior. Animal cells are often mobile,
particularly when they participate in embryogenesis. Plant cells, surrounded by a cellulosic wall, never move. There are several exceptions, however, mentioned in the following paragraphs.

Plant and animal cells do not obtain nourishment in the same
way. Animal cells are often able, like an amoeba, to obtain food by
ingesting relatively large particles: food fragments, microorganisms,
cellular debris, etc. This ingestion, or phagocytosis, occurs by forming a vesicle that surrounds the prey, followed by digestion (Figure
45). Plant cells, surrounded by a wall against which they press
through turgor, are incapable of phagocytosis. They obtain nourishment in a different manner, photosynthesis, through the activity
of chloroplasts using solar energy, atmospheric carbon dioxide, and
a mineral solution raised from the soil. The plant cell creates its food
inside itself. In the final analysis, plant and animal cells require the
same elements (carbon, nitrogen, sulfur, oxygen, hydrogen, phos-

phorus, etc.), which come from the nonliving environment in the case of plants and from preexisting organisms in the case of animals. The two cell types get rid of wastes by a sort of defecation, exocytosis (Figure 45). In plants, the cellulosic wall is formed by the cell that it contains and protects; this is a constructive defecation, of which plants provide other examples.

Simple cellular division, mitosis, is different in plants and animals (Figure 46). The animal cell is constricted by a contractile ring, separating the two daughter cells, whereas in plants a wall is formed, pierced by plasmodesmata through which the cytoplasms communicate, and the daughter cells are thus never completely separated. This has profound implications.

In classical cell theory of the mid-19th century, the cell is the basic element whose aggregation provides the final form, like constructing a wall by adding layers of bricks. Kaplan and Hagemann (1991) have developed a different concept of cells and organisms, extremely novel, in which the cell is not the basis for morphogenesis; it is only a marker. In this hypothesis the basic element is the form of the organism, secondarily subdivided, and only imperfectly, into cells. The plant is thus comparable to a house that can be subdivided into rooms with no influence on the exterior architecture. To support what they call the organismal theory, tracing the idea back to Sharp (1926), the authors use the observations of Haber et al. (1961). Wheat mutants are known that are not composed of cells, a mutation corresponding to blockage of DNA synthesis and thus loss of the mechanism of mitosis. That these mutants resemble wild wheat, albeit a little sickly, demonstrates that the form of a plant does not depend on its cells. Kaplan and Hagemann also recall that the sophisticated forms of the algae *Bryopsis* and *Caulerpa* do not rely on cellular division because those plants are not composed of cells (Figure 47). They are made of a common cytoplasm, containing numerous nuclei, and have even been considered giant single cells. Kaplan (1992) has suggested that the cell theory only applies to animals whereas the organismal theory is better adapted for plants. A subject of such fundamental importance certainly merits more research.

Plant and animal cells diverge in their capacity for differentiation, an essential cellular mechanism that is rather poorly understood. In a very young animal embryo, a little after the initial segmentation of the egg, the cells are all identical. This equivalence

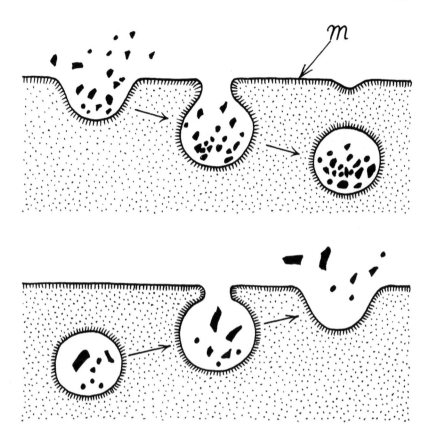

Figure 45. How cells feed themselves and get rid of waste. Above, a typical animal cell mechanism, phagocytosis. Animal cells are capable of nourishing themselves by ingesting food particles through a phagocytic vesicle. Below, a method common to the two kingdoms, exocytosis, allowing cells to get rid of wastes and, in plants, to construct the cell wall (m, plasma membrane).

does not last. A little later in its development, these embryonic cells differentiate. At first, the differentiation is subtle, mainly in the size of the cell or the amount of reserves it contains. Once begun, the differentiation process leads to the full range of extremely specialized cell types found in the adult animal.

How does differentiation occur, resulting in cells differing in so many ways as a red blood cell and a neuron, a pollen grain and a wood fiber? It is necessary to recognize that the fundamental mechanism remains poorly understood; we stand before differentiation as

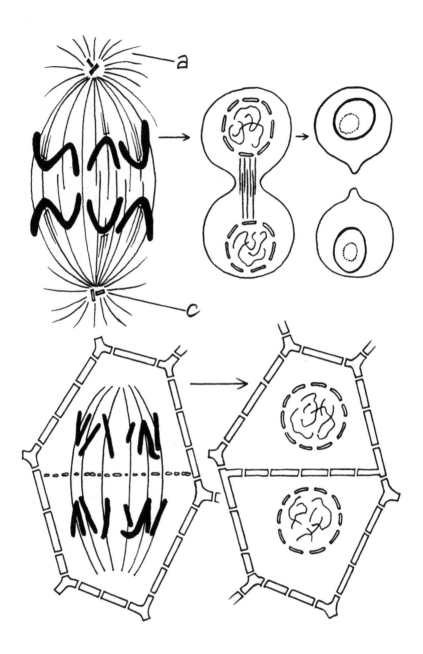

Figure 46. Cell division in animals (above) and plants (below). Asters (a) and centrioles (c) are typical of animal cell mitosis. A contractile ring splits the cell into two daughter cells, which become completely separated. In plant cell mitosis, a cellulosic wall forms, pierced by plasmodesmata through which the two cytoplasms connect; the daughter cells are thus never completely separated (Kaplan 1992).

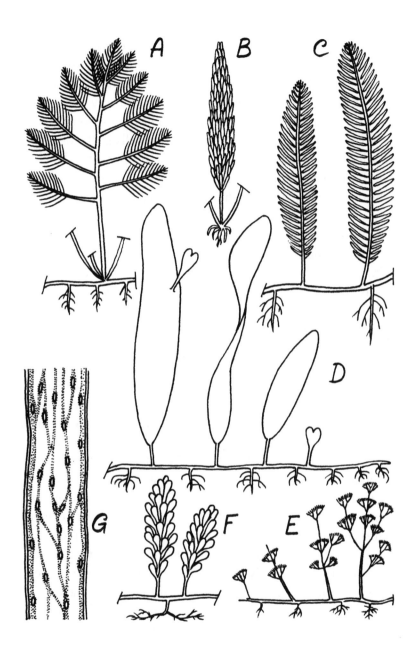

Figure 47. *Bryopsis* and *Caulerpa*, green algae with morphogenesis that does not depend on cell division: *B. plumosa* (A), *B. muscosa* (B), *C. sertularioides* (C), *C. prolifera* (D), *C. verticillata* (E), *C. racemosa* (F). Branches, roots, and leaves of quite varied form are produced with absolutely no morphogenetic role filled by cell division. (G) These plants are constituted of only a single cell with many nuclei (Kaplan and Hagemann 1991).

Claude Bernard stood before respiration. Nearly everything remains to be studied, and the difficulty comes from the paradox that the final constitution of a differentiated cell and its function seem to depend very little on the genetic information and very much on the repression or activation of certain genes. Differentiation is a function, for example, of the places occupied by cells in the organism. The selective regulation of gene expression remains to be elucidated. Biologists admit that the elucidation of these mechanisms will be their principal task in the 21st century.

My aim here is comparison. Is cellular differentiation in plants identical to that in animals? Perhaps it is at the level of the genetic mechanisms, but the two differ greatly in the magnitude of the final result. A flowering plant is composed of only 30 or so types of cells whereas a vertebrate produces at least 200 (humans, 210). A difference more important than the number of cell types is reversibility versus irreversibility of differentiation. In animals, it is unusual for cellular differentiation to be reversible. Fragments or isolated cells of sponges and some worms can reestablish a complete organism through somatic embryogenesis (Bierne 1994). But as a general rule, animal cells cannot achieve such a feat. Even if they are placed in culture, or if partial differentiation occurs, animal cells are incapable of having their development reprogrammed. This is notably true for vertebrates, and bad luck for poor Dolly, the cloned sheep whose physiological age is apparently older than it would have been thought to be chronologically, showing the difficulties of animal cloning.

Certainly, no one can be certain that this situation will not change. Progress in cellular biology will perhaps lead to a modification of this viewpoint or to its abandonment. On the basis of present knowledge, however, we can only say that the animal cell loses totipotency in the course of differentiation. The only totipotent animal cells are, in order of their appearance, the female gamete, the fertilized egg, and the first cells resulting from the division of this egg.

Where the Horticulturist Precedes the Biologist

In plant cells, the situation is very different. The common practice of stem cuttings, practiced on an agronomic scale to multiply useful plants such as poplar *(Populus)* or cassava *(Manihot esculenta)*, shows that reversibility of cellular differentiation is possible, complete, and often easy to accomplish. In a stem placed in the ground (Figure 48), cells near the two ends of the cutting reestablish mitotic activity and give birth to progenitors of roots and stems while adjacent cells redifferentiate as vessels, feeding the newly formed organs. An entire plant is quickly reconstituted.

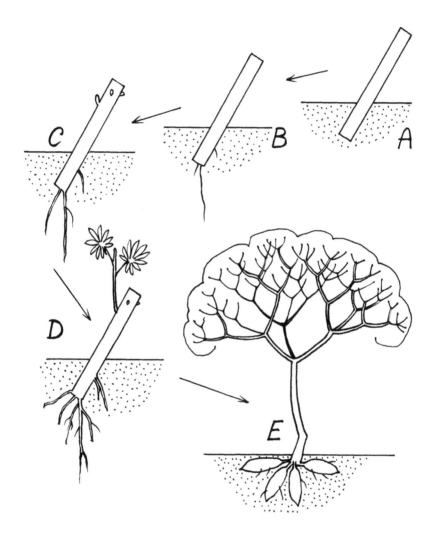

Figure 48. Propagation of tapioca, *Manihot esculenta*. (A) A stem cutting is placed in an inclined position in sandy soil. (B) Roots appear. (C) Buds on the upper part of the stem begin to enlarge. (D) One of these buds, better situated than the others, produces a leafy branch. (E) The entire plant reconstitutes itself in 4–6 months (E). All this is made possible by the reversibility of cellular differentiation.

The techniques of tissue culture clearly demonstrate the totipotency of plant cells. After cells are established in culture, modifications to the medium (including changing the concentrations of gases in the atmosphere, or of minerals or growth regulators) promote production of roots and shoots, regenerating the entire plant. Hundreds of plant species—gymnosperms, monocots, dicots—are routinely multiplied in this fashion.

In 1902, the Austrian physiologist Gottlieb Haberlandt wrote, "Placing vegetative cells of higher plants in culture in a nutritive medium will yield rich lessons on the capacities that cells possess as simple organisms" (Jullien 1980). A half century was needed for this prediction to be verified, when experiments demonstrated the reality of totipotency—controlling doses of growth regulators precisely, using isolated plant cells on various substrates to control plant differentiation. This research led to proof that plant cells can dedifferentiate completely, making them lose the memory of their prior state and leading them back to the condition of the egg. The cells redevelop like eggs, reproducing the original plant. This somatic embryogenesis has been achieved using cells of various organs, coming from plants as varied as pines, ginseng, orchids, coffee, or tobacco. It is now well established that plant cells are totipotent (Margara 1982).

Cells of certain plants and organs do not dedifferentiate: some secretory cells, guard cells of stomata, sieve tube companion cells, pith cells, root cells (Nicole Ferrière and Paul Barnola, personal communication). It is possible that this is simply a question of technical progress; it seems to me that such exceptions do not invalidate the principle of totipotency in plant cells.

Foreshadowing at the Cellular Level

This comparison between animal and plant cells can be reduced to a few simple ideas without distorting the facts. The first is that the cells are profoundly different. Plant cells are large, immobile, and self-sufficient in an energetic sense, thanks to the chlorophyll that allows them to use solar energy. They allow only small molecules to enter, and they do not release their waste products, instead storing them internally or recycling them. Unspecialized plant cells are capable, under appropriate conditions, of losing and then forgetting their differentiated states, expressing their totipotency, and reproducing an entire organism. Animal cells are small, often mobile, lack plastids and thus are incapable of using solar energy. They must nourish themselves with prey that they ingest by phagocytosis, then release their wastes. Animal cells differentiate irreversibly; they cannot, at least at present, express the totipotency of the egg cell.

Plant cells resemble plants, and animal cells resemble animals. This statement has a precise meaning, that the two types of cells impose their own personalities on those complex edifices, plants and animals. As always, however, the whole is greater than the sum of its parts. Emergent properties arise from complex cellular associations, which the knowledge of cells alone does not allow us to predict. Next, the points at which these emergent properties differ between the two kingdoms are examined.

CHAPTER 4

Plant Biochemistry in a Nutshell

Perfumes! Some fresh and cool, like babies' skin,
Mellow as oboes, green as meadows; some
Rich and exultant, decadent as sin,

Infinite in expanse—like benzoin gum,
Incense and amber, musk and benjamin—
Sing flesh's bliss, and soul's delight therein.

CHARLES BAUDELAIRE, *Correspondences*
from *Les Fleurs du Mal*, 1861

The magical flora is so vast that it suffices to push open the gate
of the garden to encounter the simplest offers, nearly transparent, in all cases honest, as well as proposals that make one blush
or scared.

PIERRE LIEUTAGHI, *La Plante Compagne*, 1991

Have not plants served humans throughout our history just as
they have served insects? Does the plant serve the gardener or
does the gardener serve the plant?

CLAUDE GUDIN, *Nique Ta Botanique*, 1996

As plants evolved they were faced with problems of waste disposal, and it is thought that many secondary plant metabolites
may have been formed as part of a sort of chemical waste disposal system . . . ; lignin formation may be the result of mutations which brought about the polymerization reactions of phenolic waste materials and the deposition of this material in the
cell wall matrix. This in turn conferred greater structural rigidity which was advantageous for the development of large terrestrial plants. Thus a secondary metabolite became a vital part
of the plant and no longer just an inconvenient waste product.

J. R. L. WALKER, *The Biology of Plant Phenolics*, 1975

M AY 1997. I see a portrait of Juliette Binoche stuck on the wall
of my town, an advertisement for the perfume Poême de Lan-
côme. I usually find these ads stupid, but this one has a profoundly
vegetable dimension that touches me. Juliette's radiant face, a small
gilded vial, and between the two, "Everything said without a word."
Beauty, silence, perfume: How better to express, apart from poetry,
the true style of plants compared to animals? All said without a word,
and I add, all done without a gesture. *Voilà*, a worthy introduction to
the biochemistry of plants.

Before entering into the heart of this subject, I wish to acknowledge the remarkable book by John King (1997), a physiologist at the
University of Saskatchewan, *Reaching for the Sun: How Plants Work.*
The biochemistries of plants and animals are marked by some significant differences. In certain situations the differences are smaller
than one would think, whereas in others they are much more profound than generally believed. At the level of the fundamental materials of life—water, proteins, lipids, sugars—the differences are not
very important. Rather, the differences are only quantitative. It is
natural to see similarities at this level because animals consume
plants, unless they eat other animals that eat plants. That the eaters
would adopt the chemistry of the eaten is not surprising, even more
so because they take in the same water and respire out the same air.
However, at this elementary level some quantitative differences can
already be observed.

The Silhouette, Cellulose or Protein

Plants photosynthesize enormous quantities of sugars, or carbohydrates. Cellulose is one such carbohydrate; it constitutes most of the
wall of plant cells and is the most common of all the organic molecules on Earth, representing more than half of all terrestrial biomass (Chapters 1 and 3). On the whole, the silhouette of a plant is
the result of the cellulose that envelops its cells and the water that
fills them.

In contrast, plants have considerable difficulty appropriating
nitrogen, which they need to synthesize proteins. Not that nitro-

gen is rare—it composes some 80% of the atmosphere and is found in enormous subterranean reserves. Plants can only use nitrogen as nitrate or ammonium, fabricated by soil bacteria and rapidly leached away by rainwater. To assure a sufficient supply of nitrogen, plants such as *Cycas*, legumes, and *Casuarina* (NRC 1984) have evolved symbiotic complexes with nitrogen-fixing bacteria on their roots. Other plants, such as bladderworts *(Utricularia)*, pitcher plants *(Nepenthes)*, and Venus flytraps *(Dionaea)*, became carnivorous, obtaining nitrogen from their animal prey.

Although they easily assimilate carbon, nitrogen remains a constant problem for plants. It is the reason that they are poor in proteins and rich in carbohydrates. The situation is the opposite in animals, which are relatively poor in sugars but rich in proteins. They eat plants or plant-eating animals. It seems paradoxical that animals become rich in proteins while eating plants in which these molecules are rare, but the quantities consumed are enormous. Proteins give the animal its overall form, going into the chitin of the exoskeleton of an insect or the epidermal and muscular tissues of a vertebrate. These differences have practical consequences, some frankly unpleasant. Animals can, to us, smell bad. First, they produce feces; I certainly do not like that smell. Paul Valéry (1950) asked, "What does the tiger make, of the quickness and elegance of an antelope? He reduces it into a tiger and into excrement." Most animals eliminate wastes through organs topographically close to those of sexuality, when it is not the same organ that performs both functions. A lack of taste, my friend Olivier Pascal affirms. I do not know how to reply to him; I fear that he makes some sense.

In any case, there is nothing like that among plants. Do they produce excrement? This question is not a simple one. They seem not to produce it. *A priori*, however, all machines, biological or not, must produce some waste. It has been suggested that plant excrement is represented by dead leaves, self-pruned branches, or even the oxygen liberated in photosynthesis. It seems that it is in biochemistry that this question can find the most satisfying answer. According to Neish (1965), plant excrement may be represented by phenolic compounds. In their free form these compounds would be toxic, and plants get rid of them by stocking them in their walls and vessels, by polymerizing them into an inert form, lignin. As vascular cells die, this storage has the double advantage of being inoffensive (Charles

Hébant, personal communication) and conferring on plants the mechanical strength that allows them to attain very large size.

An animal that stores excrement would also be capable of rising very high (Figure 49). If the hypothesis of Neish and Hébant is substantiated, we should admire plants for recycling toxic waste to produce one of the keys to their success. In this hypothesis, *déchets* (wastes) are replaced by *"inchets"* (inches) in plants, according to the astute neologism of Rumelhart (2000), and excrements by increments, growth in girth (Vincent Savolainen, personal communication). One thing is certain: The botanical solution to the problem of excretion has the merit of no odor.

A Regrettable Inelegance

We notice that animal odors are frequently repulsive. Alas, those of humans are not much better: odors of filth, of the sewer, of urine and vomit, "belches, intestinal gas, farts, colic, fetid diarrhea" (Corbin 1986). It is when they are dead and the cadavers decompose that animals release truly atrocious odors, however, those of putrefaction. It is impossible to imagine a spectacle more uniquely animal, or more totally unclean, than carrion crawling with maggots. It is impossible to imagine an odor worse than that of a rendering works. All this is the result of the high concentration of protein in animals, whose decomposition liberates nitrogenous and sulfurous compounds with names evoking their stench: putrescine, cadaverine, etc. A stench for us but not for flies, which goes to show the subjectivity of smell; no odor is good or bad in itself. To take this further, no volatile emission has an odor by itself; there must be a smeller. The human point of view, and more generally the animal one, is a fear of odor of excrement and cadavers, which has an adaptational value. The danger of infection is real, better to stay away.

Plants, on the whole, are clearly more admirable; there is practically nothing malodorous about them. Even when they are dead or decomposing, the perfumes they release are more melancholy than aggressive. It is perfectly possible to become attracted to the death odor of plants, the smell of trees being felled or of sawmills, the fragrance of timber or the hulls of old boats, the scent of dead leaves, rotten wood, humus, or the forest understory. The explanation is biochemical. Decomposition of carbohydrates, mainly

Figure 49. Solutions to the problem of excretion. An animal that stored its
excrement would also be capable of becoming very tall.

cellulose, produces no malodorous molecules. Exceptions exist but are easy to explain. Some legumes, rich in proteins, and cabbage and eggplant, smell bad when they decompose. There are also some flowers, about which I write later, that stink to us but of which flies are fond.

Incontestably, plants are talented at producing agreeable odors: the tang of a field of algae at low tide, the freshness that reigns in a pine forest, the perfume of freshly cut hay, the scent of ripe rice on the terraces of the Asian countryside, the aroma of freshly brewed coffee, the fragrance of an old garden in the first drops of summer rain. Even Georges-Louis Leclerc de Buffon (1707–1788) fell under such a charm: "A light air whose freshness I sense carries perfumes that cause an intimate blossoming and give a feeling of love for myself" (Buffon 1971).

Certainly, more than basic chemistry is involved because it is not only proteins, lipids, or sugars that produce these radiant scents. Plants differ from animals above this primary chemical level and have a second level of biochemical activity. They produce secondary chemical compounds by the thousands. Plants have at this secondary level a prodigious range of biochemical mechanisms that animals lack. Plants are virtuosos of biochemistry.

Why is there such inordinate inventiveness in the biochemistry of plants? Why must they produce innumerable secondary compounds that animals have mostly ignored? Why do we often speak of medicinal plants whereas medicinal animals—Spanish fly or leeches—seem nearly nonexistent? The answer resides entirely in the choice already discussed: to be mobile or fixed in place.

An animal, if attacked by a predator, looks for its salvation through escape. If it covets a prey, it pursues it. If it desires a sexual partner, it approaches it. If it is too cold for its comfort, it buries itself or migrates. If it is too hot, it looks for shade. In short, an animal can resort to mobility to solve most of its problems. Since plants cannot move in a desired direction or flee that which would do them harm, they must find mechanisms that compensate for their immobility. Roman Kaiser (personal communication) compares their situation to that of paraplegics, obliged to develop capacities for sociability and innovation to compensate for the lack of ability to move themselves. For the most part, plant inventions are biochemical. They are linked to respiration.

Since the time that Claude Bernard wrote (1878–79), we have known that respiration is a process common to both plants and animals. It can be summarized as follows: The mitochondria burn glucose in the presence of oxygen, liberating water, carbon dioxide, and energy in the form of adenosine triphosphate, which is used directly by the organism. This description conveys the essentials of the phenomenon of respiration that we experience as animals. Fatigued by physical effort, to regain our energy we must breathe deeply to take in oxygen and expel carbon dioxide. We also understand that if such effort must continue, a quick intake of sugar is beneficial. It can be imagined that plants must respire in a manner a little different from us because they are unaware of physical effort!

A Look at the Krebs Cycle

Because respiration must work at ordinary temperatures and not damage the cell, the combustion of glucose is a complex process involving dozens of stages, each of which results in the production of intermediate compounds before leading to the next stage, until the energy in the sugar is liberated. It is here where there are several notable differences in respiration between animals and plants.

Apart from burning sugars to furnish themselves with energy, animals must also consume proteins and fats in adverse situations. That is why the hunger strikers among us, or the landless peasants of the Third World, waste away. King (1997) wrote that beyond the fact that plants know how to breathe without lungs, they also know how to dissect the some 50 steps that compose the respiratory pathway, the Krebs cycle. At each step, plants know how to reroute a portion of the intermediate compound, use it temporarily in a parallel biochemical activity that has nothing to do with respiration, then return it back to the respiratory chain. Just as water is taken from a river, used for many purposes, then returned to resume its journey to the ocean, intermediate compounds in the respiratory cycle are used to form thousands of plant compounds. It is here, in the expression of an unbeatable biochemical virtuosity, that plants find ways to compensate for their lack of mobility. To give an idea of the extreme diversity of secondary compounds that plants produce, and the complexity of the functions they assume, I divide them into four categories: those whose function is unknown, those functioning under normal conditions, those providing relief under adverse conditions, those used to exploit animals.

THE FIRST CATEGORY of secondary compounds groups together substances whose function is unknown. It may seem paradoxical but this category is the largest. In tropical forests, when trying to identify a tree, one routinely smells freshly crushed leaves and makes a gash with a machete at the base of the trunk or on an exposed root to smell the wounded tissue and observe the sap or latex. Certainly, we obtain clues for identification, but at the same time we are struck dizzy by the abyss of our collective ignorance. What purpose is served by the volatile chemicals produced in the trunks of these tropical trees? They are responsible for such bizarre odors that we stretch our imaginations to describe them. Some smell like onions, others like grated carrots, butter, vomit, or dry sausage. To find suitable comparisons requires some mental effort, something that the humid heat of the forest does not foster: the interior of an old house freshly replastered, the cushions of a brand-new car, or urine the morning after a meal containing asparagus. The root tissues of *Philodendron* smell of orange marmalade served at breakfast in a luxury hotel. What good are all these perfumes released from the wound in a tree?

For what purpose is the red exudation, like blood, that indicates the nutmeg family (Myristicaceae)? Why the sulfur yellow latex that helps us identify members of the Clusiaceae such as mangosteen *(Garcinia mangostana)*, *Platonia*, and *Allanblackia*? Why the white latex, as abundant as milk, that points to the Moraceae, Apocynaceae, Sapotaceae, and a goodly number of Euphorbiaceae. The latter family includes the famous rubber tree *(Hevea brasiliensis)*, cultivated all along the equator and providing a living for nearly 50 million people. Despite the 6 million tons of natural rubber produced each year, we do not know why the trees exude this white latex. Can we imagine that it serves no function in the plant producing it? Certainly not—a rubber tree emptied of all its latex will certainly die.

Biochemistry for Normal Life

A second category of secondary compounds groups together those that permit plant functioning under the normal conditions of daily life. So as not to diverge from the purpose of this book, and because many others have written about the subject better than I can, I will be brief on the biochemistry of normal plant functioning. I mention first the most important of plant molecules, chlorophyll, the green pigment capable of capturing sunlight, absorbing blue to red wavelengths to make photosynthesis possible. Plants and animals share a fundamental sensitivity to light but use it in different directions: outward vision in animals and inward photosynthesis in plants (Paul Barnola, personal communication). Photosynthesis, on which life on Earth depends, is according to King (1997) the most important of all biological processes.

If photosynthesis is so important, why do animals not use it? Why is the chlorophyll molecule synthesized only by plants and a few protists? Given that it has not yet been answered, this question appears to be a legitimate and important one if I judge from the renown of the scientists who have discussed it. Pierre-Paul Grassé (1973) mused, "Why has evolution not produced half-plant and half-animal beings? We can easily envisage a worm exposing on its back a long strip of transparent cells filled with chloroplasts and producing sugars, proteins, all made from simple materials: water, carbon dioxide, nitrates, nitrites, oxygen. . . . A fish furnished with filamentous tassels crammed with chloroplasts and able to absorb nitrates dissolved in the surface waters through which light penetrates would have no need for a digestive tract. Moving and sensitive, it would feed itself like an alga. But this mixed creature, logically conceivable, would it be able to unite two such irreconcilable qualities? It is right to ask this. Mobility exacts a considerable amount of energy, usable in an instant. For photosynthesis to provide adequate nourishment for an organism of modest size, a very large surface directly exposed to light is required. . . . Such a chlorophyllous structure, from its mass and bulk, makes rapid and coordinated movement difficult, even impossible."

For Grassé, mobility excludes the presence of chlorophyll. It seems that the reverse is true for Stephen Jay Gould, who sees the presence of chlorophyll as excluding mobility for reasons of evolutionary contingency. Gould (1987) imagined that plants could manage better if they were able to move from shade toward sunlight but that inherited structural constraints never permitted making an attempt toward this fascinating option.

Please forgive my entering into a debate launched by illustrious zoologists. If I imagine myself endowed with chlorophyll-containing ears, it is to indicate that despite the arguments of my predecessors, the question about the lack of photosynthesis in animals does not seem to have gotten a definitive answer. After all, if a glider can carry solar panels to power its flight, why would not some animal do as well? It is possible to imagine various ways for reconciling humans and photosynthesis (Figure 50).

[Some marine gastropods, sea slugs, have evolved a symbiotic association with the plastids of certain siphonous green algae (Figure 47; Rumpho et al. 2000). These animals look very much like leaves floating in marine coastal waters, complete with veins. The sea slugs consume algae and digest the cytoplasm but take up the chloroplasts into their digestive cells. These cells are distributed throughout the digestive system immediately beneath the surface of the animal, giving the entire slug a green appearance. These chloroplasts, "kleptoplasts," remain active for many months, fixing carbon for the host animals. The sea slugs may obtain more than half their carbon this way (Raven et al. 2001). This long association raises suspicions about the possibility of horizontal gene transfer from the algae to the sea slugs (Douglas 1998). —translator's comment]

After chlorophyll, I mention phytochrome, a second exclusively plant pigment, blue and capable of absorbing the red and far-red wavelengths. Only discovered in the 1960s, much more recently than chlorophyll, it is nonetheless of nearly equal importance since practically all plants, from the most ancient to the most advanced, contain it. Light controls the life of plants, and it is phytochrome that allows plants to keep in touch with their light environment, down to the smallest details. Germination of seeds, growth of seedlings, and formation of flowers are three major events influenced by phytochrome (King 1997).

Phytochrome is what orients growing young shoots toward light and controls the folding of leaves of many plants at dusk. In August, it also informs the trees of the Northern Hemisphere about the lengthening nights, letting them know that winter is approaching and that it is time to form protective buds while temperatures remain moderate. Chlorophyll and phytochrome, a pair of accomplices, allow complete exploitation of the luminous universe surrounding plants.

Figure 50. Some attempts to reconcile human beings with photosynthesis.
Above, two scenes taken from *L'Histoire de l'Homme Goémon* (Dauchez 1900).
We see Captain Mateur and his crew discovering, on a desert coast, a man
covered with seaweed who later is revealed to be ... the devil! At the lower
left, a green man from the cornice of a column preserved at the Museum of
Archeology, Istanbul, Christian art of the sixth century. Green people, symbols
of our unity with nature, have an important place in European mythology.
At the lower right, the author as seen by Richard Palomino.

Growth regulators, which had first been viewed as hormones (Chapter 2), should be added to the secondary compounds plants use under normal conditions. We now know that they are not hormones and that there are several hundred of them. All organs, all plants, seem to contain these indispensable growth substances. Auxin is the best known; it mediates the curvature of shoots toward light, stimulates the growth of roots, and inhibits the growth of lower branches in young trees, giving the latter their lollipop form, pointing toward the sky. Gibberellins stimulate seed germination as well as shoot elongation. Cytokinins stimulate production of new cells. Abscisic acid can suppress the growth of shoots, and ethylene accelerates the maturation of many fruits.

Biochemistry for Relief

A third category of secondary compounds groups together those that allow plants to survive under adverse conditions or to repel predators as a substitute for being able to flee. In August in the Northern Hemisphere, when phytochrome is telling plants about the lengthening nights, it is still warm but the plants are already making preparations for the coming winter. Not able to flee at the departure of the sun, as a swift or stork can, the plants prepare to sit through the winter by making use of their biochemical resources. Abscisic acid is in charge of preparation for winter. It slows growth and controls the formation of bud scales destined to protect the most important and also the most fragile cells, the embryonic cells that form the meristems. A little later, toward October, the plant resorbs all that can still serve its needs from its leaves, and abscisic acid intervenes again to cause fruits and dead leaves to fall. Leaf fall interrupts translocation in the phloem tissue. The plants—oak, larch, lilac, apple tree—are ready for winter.

Another area in which being fixed in place is a handicap, from my point of view at least, which is that of an animal's, is discouragement of competitors. Here again, whereas animals attack or flee, plants reveal their biochemical virtuosity. We have known for some time that walnut trees (*Juglans regia*) suppress the growth of neighboring plants within a radius of several meters, as much as 25 m for a large tree. The leaves and branches produce a toxin, juglone, that is leached by rain, enters the soil, and blocks the germination of seeds, giving the walnut exclusive use of the site's resources. We now know that many other plants eliminate competitors as the walnut does, through allelopathy. Rather than put energy into movement like animals, metabolic energy is invested by plants in the synthesis of sophisticated defensive secondary compounds.

Allelopathy is the gentle equivalent of the battle of two male apes coveting the same female, the fighting of two hyenas over a carcass, or competition in boxing or Ping-Pong between a contender and the title holder. Politicians reinvent allelopathy when they invest less energy into activities benefiting their fellow citizens, preferring venomous slander or deadly insinuation to fight their adversaries.

It is in repelling predators that plants rely most on their biochemical resources, and here we see best the differences between the defensive strategies of animals and plants. What does an animal do when threatened by a predator? It defends itself, faking its death, rolling up in a ball, fleeing, or attacking. If of a species that has evolved such a strategy, it can also avoid being attacked by mimicking something inedible, even poisonous or dangerous. There is nothing comparable in the repertoire of plants attacked by predators. The predator is always an animal; plants do not attack other plants, even if some are parasites such as mistletoe, *Rafflesia*, *Orobanche*, dodder *(Cuscuta)*, or strangler fig.

The resistance, at the same time fierce yet marked by unselfishness, of acacias against kudus *(Tragelaphus strepsiceros)* has been admirably described by Wouter van Hoven (1991) of the University of Pretoria. Kudus are robust gazelles that consume the foliage of *Acacia caffra*, a tree of South African savannas. When a hungry kudu approaches the first acacia and begins to browse its foliage, all is well at first, and it eats for several minutes. Then, well before being satiated, it leaves the first tree, moves toward a second of the same species, and continues to browse to the detriment of that individual.

Why did the kudu leave the first tree while it was still hungry? Van Hoven has shown that after several minutes, the leaves of the first tree become astringent and cease to be palatable to the kudu. Analysis of the leaves of the first tree reveals that they quickly increase the concentration of tannin, a phenolic compound with the astringent flavor. Van Hoven made another extremely interesting observation. The second acacia is in a particular relationship to the first tree. The kudu goes upwind, traveling from the first to the sec-

ond tree. His analysis confirmed that all the acacias downwind of
the first tree become astringent without having been attacked. The
first tree warns all the trees downwind of the arrival of a kudu, and
the kudu knows this. Van Hoven showed that the message moving
through the atmosphere between the acacias is ethylene, a gas that
the plants produce when they are wounded and that also has a role
as a growth substance. Its movement downwind explains how the
trees upwind remain edible.

An Altruistic Tree

Two comments need to be made. First, tannins are not permanent
constituents of the leaves of *Acacia caffra*; they are a defense induced
by the predator. Moreover, the tannins slowly disappear after the
departure of the kudu; in several days the tree is edible again. Second, kudus can, without being inconvenienced, depend on acacias as
a regular food source. The only constraint is the need to change
trees from time to time. A tree, and it is here that its description as
altruistic comes into play, accepts light predation. The degree of
predation depends on the density of kudus; van Hoven (1991) indicated that if kudus do not number more than three per 100 hectares,
the two partners coexist. As Kipling might have said, things can continue for a long time in peace, my dear, on the savannas of South
Africa.

Alas, that calculation does not include the farmers who cut the
savannas into ranches of varying size, fencing them off with barbed
wire. The first dead kudus were noticed after 1980. Their condition
seemed inexplicable; they were excessively gaunt and visibly starved,
but autopsies revealed that their bellies were full of acacia leaves.
Van Hoven resolved this apparent paradox with an illuminating
explanation. Not able to move upwind because of the fences, kudus
trapped on a ranch had to eat astringent foliage because the acacias
there had insufficient time to return to edibility. Even though their
bellies were stuffed with leaves, the kudus were not nourished
because tannins inhibit digestion of proteins. The kudus died of star-
vation, stomachs full, killed by acacias whose charity turned into
lethality. Although they tolerate a little browsing, the trees know
how to deal with predators who press too hard. Kudus continued to
die back to the normal density of three per 100 hectares found out-

side the ranches. The most astonishing thing, van Hoven reported, is that the acacias help the kudus, animals of little use to them, survive. Astonishing, yet it is common for plants to provide services to those who can do them wrong. Siddhārtha Gautama, the Buddha, was astonished that the forest, with gentleness and unlimited good will, offered its shade even to the woodcutters who destroy it.

The example of passionflowers *(Passiflora)* and *Heliconius* butterflies deserves mention here. Despite the fact that herbivores succeed in circumventing the chemical defenses of the plant, the plants retain an advantage. These surprising observations have been accumulated by Larry Gilbert (1975) of the University of Texas at Austin. Lianas of the American Tropics, passionflowers are a food source for the caterpillars of the butterflies (Figure 51). Since their leaves contain alkaloids and cyanogenic glycosides, the lianas are toxic to most predators, but the *Heliconius* caterpillars have found a way to feed on the leaves avoided by other herbivores. The caterpillars, nourished by tissues containing toxic secondary compounds, become toxic themselves. In turn, they metamorphose into butterflies that are equally toxic. As the *Heliconius* butterflies are brilliantly colored, birds recognize and avoid them. It is thus that secondary compounds produced for the protection of a plant also protect a butterfly!

A Butterfly That Remembers Shapes

Gilbert (1991) showed experimentally that *Heliconius* butterflies are able to see and remember shapes. They are able to recognize the species of *Passiflora* on which they must lay their eggs by the shape of the leaves. Over evolutionary time, a sort of equilibrium has been established in which each species of *Heliconius* has adopted a single species of *Passiflora* to feed its caterpillars.

To establish an appropriate shape, it is as if the passionflowers were conscious of the ability of the butterflies to recognize their leaves, and that they consequently tended to modify the form of the leaves as much as possible. This allowed the plants to reduce, or even avoid, predation by the caterpillars and attain sexual maturity. Gilbert has shown that too much predation on the foliage of a passionflower prevents it from flowering. A genetically anomalous passionflower, whose leaves diverge from the norm of the species, will

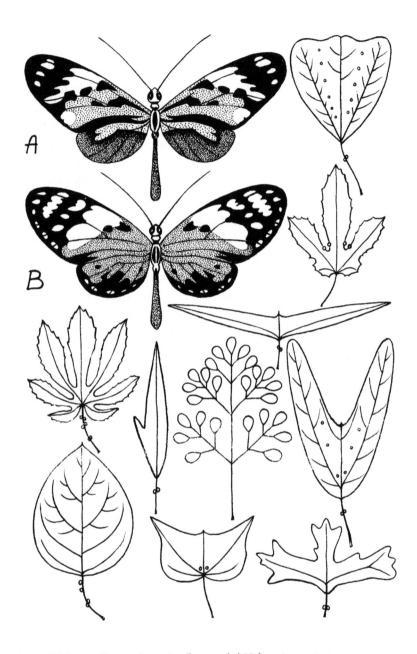

Figure 51. Butterflies and passionflowers. (A) *Heliconius numata* (Heliconiidae), a superb South American butterfly with brick red, black, and white coloration. It is toxic and birds avoid attacking it. (B) *Melinaea ludovica* (Ithomiidae), a mimic, living in the same area but not toxic. This butterfly benefits from its resemblance to *Heliconius;* birds avoid it. The other drawings illustrate passionflower leaves. *Passiflora* is considered to have the greatest diversity of foliage form of all flowering plants.

increase its chances of avoiding predation and thus increase its chances of reproducing. The more different the leaf, the more secure the plant. The exceptional variation in leaf form among passionflowers is thus quite understandable (Figure 51). This diversification of foliage could lead in turn to speciation; the passionflowers have diversified under the influence of butterflies in the same way that the butterflies have evolved under the influence of the biochemistry of the passionflowers.

So, passionflowers defend themselves by the toxic secondary compounds that accumulate in their tissues, but such biochemistry empowers plants to achieve much more subtle modes of defense (King 1997). Substances that mimic insect hormones occur in massive doses in the tissues of certain plants: ferns, conifers, rushes. These pseudohormones make the plant-eating insects that consume them sterile, reducing the population of insects, thus reducing herbivory. Other plants have found an original way to defend themselves against caterpillars; they emit terpenes that have a hormonal action on the insects. The caterpillars become incapable of metamorphosis, they cannot reproduce, and herbivory is reduced.

Certain plants dissuade insects with means even more sophisticated. Some aphids emit a gas as a chemical signal when attacked by a predator. This alarm pheromone alerts other aphids, signaling flight. A plant of the nightshade family (Solanaceae) produces this same alarm pheromone. An attack by aphids releases the gas, making them move away. The plant repulses a pest by mimicking the pest's alarm signal, a nearly perfect deterrent (King 1997). Fixed in place as they are, plants repulse predators through numerous means. Moreover, the gems of their biochemical arsenal are deployed not to dissuade animals that harm them but to attract those whom they need.

Biochemistry to Take Advantage of Animal Mobility

The fourth and last category of secondary compounds groups together those that let plants borrow the mobility of animals to complete their own sexual reproduction or to disperse their descendents. Many plants—algae, mosses, ferns, lycopods, pines—have no need of animals, with whom they maintain relationships involving only conflict: herbivory, predation, biochemical retorts, etc. It is wind (or marine currents in the case of algae) that carries spores, gametes, or pollen, and it is wind that disperses the winged seeds of conifers. Plants belonging to the ancient groups—algae, bryophytes, ferns, early seed plants, gymnosperms—can generally do without animals. To be more precise, they have no need for animal mobility.

Yet even among ancient plants, animals began to play a useful role. I know of a weevil from New Guinea that assists in the dispersal of lichens by carrying them on its elytra. I know of other interactions: a weevil that pollinates the cycad *Zamia*, rodents that disperse the seeds of *Ginkgo*, insects that pollinated the flowers of fossil Bennettitales (Taylor and Taylor 1993), flying reptiles (pterodactyls) that ate the ovules of cycads and probably assisted in their dispersal, birds that disperse the seeds of yews or pines in the Northern Hemisphere, or araucarias in Australia, and the slow loris whose rough coat shelters algae. These are a few examples. Cooperation between ancient plants and animals only sketches a rough outline. These plants did not take much advantage of animal mobility because they did not have much to offer to the animals. In these plants, the boundary between pure and simple animal predation (of pollen or seeds) and more beneficial functions (pollination or dispersal) was not clear, at least through the Jurassic period.

The situation changed at the debut of the Cretaceous period with the expansion of the angiosperms, plants with true flowers and fruits, such as grapes and olives, baobabs and sweet peas. Then, there was clear development of cooperation between plants and animals, the former contributing their biochemistry, the latter their mobility. Plants offer perfumes to seduce animals, colors and shapes to attract them, and foods to nourish them. In return, a first group of animals—bees or hummingbirds, butterflies or bats—carries pollen, facilitating sexual reproduction. A second group—ants or termites, marmosets or bats—disperses seeds. For the plant, this corresponds to an extension of its territory. After the Cretaceous, plant–animal cooperation evolved vigorously, to the extent that such mutualistic interactions evolved together with the angiosperms, whose supremacy demonstrated the success of this pooling of resources.

However, conflict between plants and animals, herbivory and predation, did not cease. Some zoologists continue to focus on the balance between predation and mutualism. Caterpillars browse leaves, bees gather pollen, and birds disperse seeds by eating fruits.

Predation and mutualism are not as distinct as we would believe; we must look beyond biology to distinguish them because, to the animal, everything is predation. An animal may provide an essential service to a plant, but it does so out of ignorance. In feeding on nectar or pollen, it pollinates; in eating a sweet fruit, it disperses seeds.

Are Animals Manipulated by Plants?

Animals know that they take but ignore what they give; the subtleness of plants occurs through their manipulation of animals, using toxic, colorful, or volatile secondary compounds. Put aside any innate feeling of condescension toward our green cousins and observe them exploit animals. Biochemical defenses, such as green flower buds indistinguishable from foliage, allow plants to camouflage parts when the presence of animals is not desired. Plants know how to attract animals when they need them, however, with a brilliantly colored corolla, sometimes a delicious fragrance, the promise of a meal of nectar.

Seeds must be scattered only when they are ripe. They need time to mature and the attention of animals must be discouraged before then. The corolla drops off and only green fruits remain, hermetically sealed, odorless, often spiny and toxic (Yves Gillon, personal communication). The seeds quietly mature inside.

Poor animal, its role is not finished; it needs to disperse the seeds to satisfy the geographic ambitions of the plant. The fruits acquire an appetizing color of ripeness, or they open and present an enticing appearance, like a pizza or an apple turnover, while beckoning aromas diffuse out, with an effect like that of roast chicken in a market. Animals gather. They believe themselves victorious, unaware that they are doing what is expected of them. For me, the image evoked is that of a duchess seated on a throne. She rings her flunky for tea and scones, dismisses him with a wave of her hand to enjoy her tranquility, calls him to remove the service and clean up the crumbs, then sends him to town to do the shopping. The guile is to change the scenario without the animal's notice. It sees itself as a predator and acts as one while it has really become an assistant, almost a flunky, sometimes a gigolo as in the case of *Ophrys*, as explained later.

"The animal is essentially distinguished by the presence of a stomach and has been well described as an animated digestive organ," an opinion expressed in 1828 by Estiènne during a meeting of the erudite Polymathic Society of Morbihan in Vannes, France. This point of view is rather unkind to animals, but not untrue. In the entire history of cooperation between plants and animals, animals have practically only one thing on their mind: Eat, eat quickly, and eat lots. For animals it is a question of life or death. Thus it is easy to control animals through their stomachs. Plants do not have such problems nourishing themselves since their food is ubiquitous. Despite their alleged superiority because of the presence of a nervous system, the perpetual hunger of animals is controlled by another, slower living being, more discreet and living at a more peaceful tempo. Animals, feverish in their search for food, are manipulated by plants whose trump card is their sidestepping of time.

Is the term manipulation excessive? Do not say yes too quickly. First, look at the effects of plant biochemical processes on animals. I cite a famous example from the British botanist E. J. H. Corner (1949), who studied in Malaya at a time when the countryside was still mostly green: "In the forest, durian trees *[Durio zibethinus]* commonly occur in groves. In the season the smell of the fruits attracts the elephants which congregate for the first choice, then come the tigers, pigs, deer, tapir, rhinoceros, monkeys, squirrels, and so on down to ants and beetles which scour the last refuse. The jungle-folk build tree shelters whence they can reach the ground when a fruit drops, and whither they can climb again to safety. Under the big trees are leaning saplings, frayed bark, trampled shrubs, and churned ground, as scenes of elephantine supremacy." How better to understand the attraction that a fruit odor has on the fauna of a forest? A durian fruit's odor is impossible to ignore and unforgettable by those who have smelled its indescribable and powerful fragrance. It is a perfume and a flavor at the same time, with unexpected notes, fecal and sexual, that provoked shocked, sometimes enthusiastic, comments from the British colonials (Alfred Russel Wallace, for example). They, along with their Malay subjects, had the urge to share the fruit with their friends, having fun, as we might enjoy sharing a melon in the evening under the arbor. The durian odor is made up of at least 26 volatile secondary compounds, the two main ones being propanethiol and ethyl α-methylbutyrate (Baldry et al. 1972).

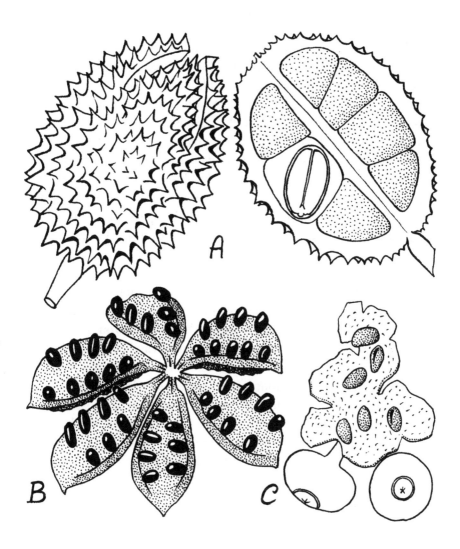

Figure 52. Fruits that whet the appetite. (A) Durian, *Durio zibethinus*
(Bombacaceae), a large tree of Asian tropical rain forests with fruits 20 cm
long. The creamy yellow aril surrounding the seeds attracts predators with
its indescribable odor (Corner 1949). (B) *Ingonia digitata* (Sterculiaceae), a
small tree of African tropical rain forests with fruits 25 cm wide. When ripe,
these fruits expose shiny black seeds on a red pericarp, an appearance very
much like that of a pizza garnished with black olives. (C) *Siparuna guianensis*
(Monimiaceae), a small Amazonian tree producing a type of fruit unknown
in Europe: a dehiscent berry. Open at maturity, this fruit (2 cm in diameter)
resembles a pastry with pine seeds.

The attraction of animals to fruits is not always olfactory; it can be visual or both olfactory and visual when colored pigments are present in addition to volatile compounds. Do we wish to bite into a peach or climb into a cherry tree? We only do what pleases us with trees that wait on us to disperse their seeds, and their biochemistry is adapted and oriented toward this goal. Tropical fruits hold the record for visual display. *Ingonia digitata*, a small African tree of the cocoa family (Sterculiaceae), displays black seeds at maturity, reflecting sunlight in brilliant points of light. Its seeds, the size of olives, are released from the red velour of its sumptuous pericarp (Figure 52). Impossible to miss, the open fruit resembles a pizza. The forest fauna treats it as one would expect.

THE TRUE VIRTUOSOS of animal attraction are not fruits, however, but flowers. Flowers! We touch on one of the pinnacles of beauty of the planet, well deserving of a blue ribbon. What would the train station at Sète be without masses of geraniums and begonias? What would Montpellier be like without its rockroses, Bormes without mimosas, Sénanque without lavender, Grasse without carnations, Pont-l'Abbé without hortensias, Singapore without orchids, Mexico City without marigolds, Japan without flowering cherries, the Alps without gentians, Valencia without oranges, England without rhododendrons, Andalusia without bougainvilleas?

A Pinnacle of Beauty

Most flowers exist to attract animals, to make a deal with those who come to receive a reward. For some animals, the hope of a reward will suffice. Next, it is necessary to cover the animal with pollen, then direct it toward other flowers to search for the same reward— or the same hope—so that it leaves pollen grains, containing the male gametes, on the receptive female structure placed there for that result. The result is that the animal memorizes the reward—or the hope of one—stimulating it to repeat the process. And the reward for the plant?—genetically advantageous cross-fertilization.

And humans? If we find that flowers are beautiful, is this because we raise to the rank of cultural values their characteristics—form, color, fragrance—that have only adaptive value in attracting animals? This would be an anthropocentric point of view; the beauty of flowers also has, in relation to us as well, an adaptational value. A flower pleases? We quickly place some seeds in our pocket, and our gardens are populated by beautiful strangers. Lilies from Iran, soybean and *Forsythia* from China, *Philadelphus* and *Gaura* from the United States, *Calceolaria* from the Andes of Peru, *Pelargonium* from South Africa—those who collected these plants from nature the first time, were they aware of what was expected of them, spreading seeds all over the world?

As expected, biochemistry provides the cues to attract animals. Anthocyanins give tints of red or blue to petals, attracting bees. If birds need to be attracted, carotenoids provide the colors of carrots: yellow-orange to red. In the small phlox family (Polemoniaceae), floral pigments vary along a north–south axis spanning the Americas to attract the most active pollinator in each region. Tropical members of the family produce fire red flowers to attract hummingbirds. Those at temperate latitudes have purple flowers, pollinated by butterflies. Finally, far north, blue flowers attract bees (King 1997).

Colors join with form. In many flowers that have a single plane of symmetry and resemble faces—*Pelargonium*, *Commelina*, orchids, *Salvia*, pansy—colored bands or lines guide the pollinator toward the nectaries, and the stamens. These nectar guides bring to mind the lines of lights that guide night arrivals at an airport.

Colors and shapes also join with fragrance, which doubtless is the pinnacle of perfection. No one knows better than Étienne Pivert de Sénancour how to explain the dizzying power of fragrance for those endowed with a refined sensitivity: "A jonquil was flowering on the supporting wall. It is the strongest expression of longing; it was the first perfume of the year. I sensed all the good will directed toward humanity" (Corbin 1986). Sénancour is unequaled; I limit myself to some plants whose particularly sumptuous odors remain in my memory after many years of research in the tropical forest. Besides tuberoses, coffee trees, and jasmines, I can add less well known species that do not have familiar common names but that are among the olfactory treasures of the Tropics: a giant liana from the Ivory Coast, *Leptactina densiflora* (Rubiaceae) with beautiful white star flowers; a hardy herb cultivated around village huts in Madagascar, *Hedychium coronarium* (Zingiberaceae), which produces large white flowers in terminal clusters; a small tree of Yunnan, *Osmanthus fragrans* (Oleaceae); and finally, and particularly unforgettable, an Amazonian tree whose red flowers are borne at the base of the trunk, *Couroupita guia-nensis* (Lecythidaceae). Familiarity, even fleeting, with these paragons of olfactory sweetness leaves an astounding feeling of well-being in our heads, an intoxicating lightness and confidence in the future. I claim that to become familiar with these perfumes is to be susceptible to a permanent change in personality and view of the world. I would do better to write only of such things for myself; the feelings are indescribable and the exaltation not expressible in words.

At the biochemical level, these odors are equally complex. The volatile oils produced in nature occur in mixtures of hundreds of different kinds of molecules: 228 compounds in the odor of an apple, 275 in that of a rose. Jaubert (1987) estimated that there are 25,000–30,000 different odor-producing molecules synthesized by plants. Among the pleasant smells, we also find a few of quite a different nature.

Nauseating Flowers

> Booz only knew that a woman was there.
> And Ruth only knew what God wanted of her.
> A fresh perfume came from the bunches of asphodel;
> The exhalations of the evening floated over Galgala.

Did Victor Hugo (1883) truly appreciate their debatable fragrance or were his asphodels only there for the rhyme? Whatever Hugo smelled, some flowers clearly do stink: some aristolochias, some arums, most species of *Stapelia* and other similar genera of the cactus-like Asclepiadaceae, finally the formidable flowers of *Rafflesia*, offering a rich olfactory range ranging from spoiled Roquefort cheese to an overflowing cesspool, sometimes a festering or, even more atrociously, rotting meat or an opened grave. Strangely, visual mimicry accompanies this olfactory mimicry, and the flowers of *Aristolochia*, *Stapelia*, and *Rafflesia* that spread this "perfume" are fleshy, somber, veined brownish red or marbled black. Flies arrive in quantity, as we would expect, showing that their senses and ours perceive similarly even if leading to opposite behaviors. This important point has been illuminated by Aline Raynal (1987), professor at the National Museum of Natural History, Paris: Bad floral odors "are connected, in our olfactory memory, most often with offensive animal odors." It is clear that when plants stink, they imitate animals.

Copulating Flowers and Animals

Not content to control animals through their stomachs, plants have succeeded in controlling them through their loins! Known by many naturalists, the story of the orchid *Ophrys* has its place here. The lips of these ravishing Mediterranean orchids resemble bees. The mimicry is advanced, involving color, form, texture, hairiness, and even odor. The lure includes chemicals identical to the pheromones produced by queen bees, complex mixtures of aliphatic compounds—alcohols, carbohydrates, etc.—and terpenoids—farnesol, geraniol, etc.—more than a hundred volatile compounds in all. Male bees attracted by *Ophrys* flowers are not looking for food.

Bertil Kullenberg, an ecologist at the University of Uppsala, has studied the sequence of events leading to a unique feat, seemingly against nature: sexual intercourse between an animal and a plant (Kullenberg and Bergstrom 1986). More than 100 m downwind, the bee detects the first volatile substances and follows the odor, flying toward the *Ophrys* flowers. At 1 m it sees the floral lip, confuses it with a female bee, and the final guidance is visual. Placing itself on the flower, it takes the position for copulation (Figure 53). Kullenberg has filmed the scene.

This really is copulation—at least as far as the insect is concerned. The bee is clearly more excited than during normal copulation and spends as long as 20 minutes in the flower, a period during which its head brushes against a sticky knob, the rostellum, which is the key to the device for releasing the pollen sacs of the orchid. When the bee leaves, it carries the pollinia, comprising millions of agglomerated pollen grains. After flying several minutes, a fresh, new copulation with another *Ophrys* flower places the pollinia on the stigma, and fertilization eventually takes place—at least as far as the orchid is concerned.

This is a strange affair, occurring only in *Ophrys* and a few other orchids. The copulation is real only for the insect but is followed by fertilization that is real only for the plant. Unaware, the bee transmits the plant's male gametes, completely squandering its own. How can we see this as other than manipulation of the bee by *Ophrys*? Certainly, this plant–animal coupling is sterile, but I cannot stop myself from dreaming what the result could be! It seems that Norman McLaren and Edward Lear have had similar dreams (Figure 54). These images, stylized and lacking in seriousness, nevertheless suggest genetic and evolutionary dilemmas.

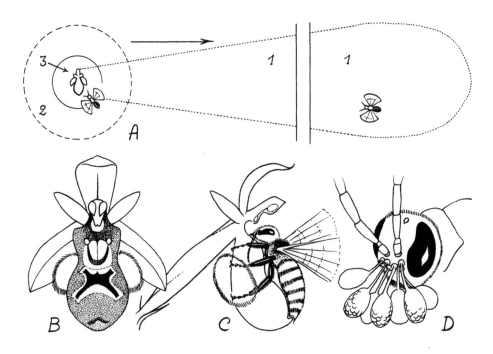

Figure 53. *Ophrys* and bees. (A) Volatile chemicals produced by flowers of the orchid *Ophrys* are distinguished by their range of action and their effect on male bees. In zone 1, more than 100 m downwind (indicated by the arrow), the bee is guided only by odor. In zone 2, a meter in diameter, the flower's appearance adds to the odor's attraction. In zone 3, only 1 cm in diameter, touch takes over from sight (Kullenberg and Bergstrom 1976). (B) a flower of the bee orchid, *Ophrys apifera*, seen head on and about 2 cm high. (C) Initial coupling of a male bee, *Eucera longicornis*, with the lip of the orchid. (D) A male bee, *Argogorytes campestris*, carrying a number of pollinia on its head, souvenirs of visits to several orchid flowers.

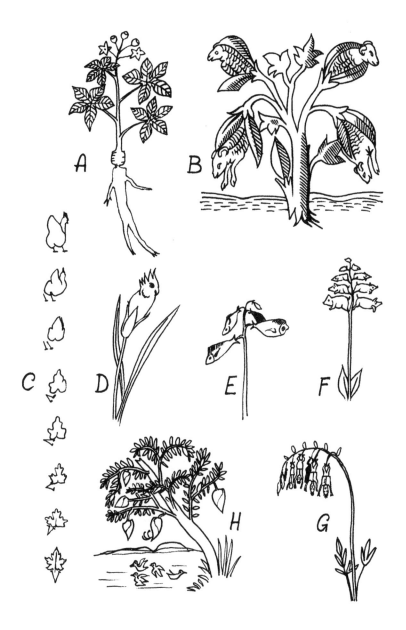

Figure 54. Ancient fantasy, a fusion of plants and animals, including humans. (A) Ginseng, "in its ambivalence of a nature half-plant, half-human," China 1552–1578. (B) The vegetable sheep of Sir John de Mandeville, 1633 (A–B, Lieutaghi 1991). (C) The goofy chicken of Norman McLaren (1947). (D–G) *Nonsense Botany* of Edward Lear (1888): *Cockatooca superba* (D), *Fishia marina* (E), *Piggiawiggia pyramidalis* (F), *Manypeeplia upsidownia* (G). (H) The willow that bears both ducks and barnacles, from Ulisse Aldrovandi (16th century; Bertin 1950).

CHAPTER 5

Evolution

The great injustice of nature is to have not known how to limit itself to a single kingdom. Compared to the plant's world, everything appears inopportune, poorly done. The sun should have ignored the arrival of the first insect, and moved elsewhere with the invasion of chimpanzees.

ÉMILE MICHEL CIORAN, *De l'Inconvénient d'Être Né*, 1973

I found a longing coming over me . . . to live up there too; such is the strength and certainty that this tree had in being a tree.

ITALO CALVINO, *The Baron in the Trees*, 1959

To you, for whom a hole in the earth is a tomb, how to explain that for me it has been a cradle?

MICHEL LUNEAU, *Paroles d'Arbre*, 1994

BIOLOGICAL EVOLUTION is a reality to which all living beings are subject, and no serious biologist would dream of questioning it. According to the classical Darwinian sequence, differences accumulate from one generation to the next, gathering strength from one geologic age to another, species changing, diverging, proliferating, splitting into new species.

Do Plants and Animals Evolve the Same Way?

If evolution is inseparable from life itself, does that mean that all beings evolve in the same fashion? Is it legitimate to ask this question? It is rarely asked. It was posed by van Steenis (1976), then by Sutherland and Watkinson (1986), and their answers were no. Since then, other differences in evolution between the two kingdoms have been proposed. It may surprise us to see differences in evolutionary mechanisms between plants and animals, but rather, should it not be astonishing that two groups so different maintain such similar fundamental processes? The genetic information conveyed by DNA and subject to mutation, sexuality with its meiosis and fertilization, selection, speciation, phylogeny, etc.—there is nothing in these that is specifically animal or plant. Are the differences trivial or do they transcend the similarities? Let us examine the facts first.

The differences are numerous. They occur at various levels, and I begin, a bit arbitrarily, with those concerning generations. What do we call a generation? If we take a moment to study the common and pretty European fern, polypody *(Polypodium vulgare)*, in front of our eyes—the actual fern—is the first generation (Figure 55). The cell nuclei of this fern contain 74 pairs of chromosomes; from a genetic perspective, the polypody contains doubled information. This generation of the polypody is diploid. Each gene may be found in two forms, termed alleles. When a plant is in a diploid state it can contain an allele on one chromosome that is not expressed while the allele on the homologous chromosome is expressed. In a similar way, a truck with double tires on its rear wheels continues to roll along if one tire bursts. The expressed allele is said to be dominant, and the other recessive. Mute as they are, recessive alleles do not contribute less to the enrichment of the genome; they are the material on which evolution works.

On the lower side of its fronds, the polypody produces sporangia, tiny organs barely visible to the eye. In them, meiosis, the special cell division reserved for sexuality, gives rise to cells in which the genetic information is unique, each chromosome only occurring singly. These cells are haploid. To be precise, these cells are spores, and the plants producing them are sporophytes. The polypody fern is a sporophyte. The spores are dispersed by wind. Falling on a shady, moist site, the spore germinates and grows into a second plant (Figure 55) very different from the polypody. A small green plate, barely visible, ephemeral and haploid, evoking little of a fern, of anything, except to a well-informed botanist. It produces male and female sexual cells, gametes; the plant is thus a gametophyte, the haploid generation of the fern. This second plant, which gardeners and greenhouse growers call the prothallus, is biologically just as important as the other, more visible generation. The latter, which anyone can observe and handle, has no possibility of existence without the tiny, discreet generation, its indispensable complement.

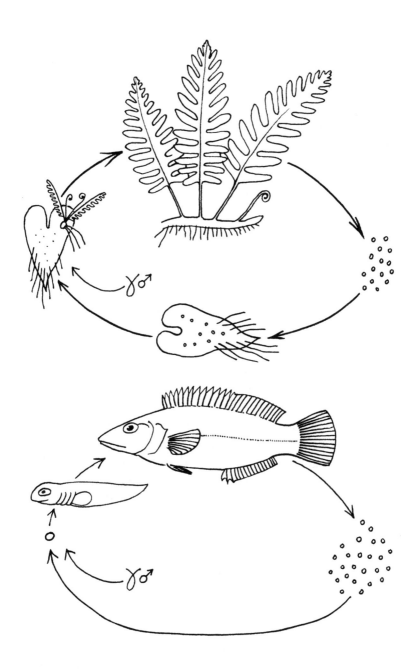

Figure 55. One generation or two. Above, in a fern, spores (on the right) give rise to a gametophyte, which is the sexual generation. The arrival of a spermatozoid (male gamete) allows a return to the original fern. One plant, two generations. Below, in a fish, the sequence of events is much simpler. Suppose that the fish is a female; it produces unfertilized eggs, and the arrival of a spermatozoid allows production of a new fish. One animal, one generation.

As a student, I had believed for a long time that these two generations were merely manners of speaking, pedagogic tricks, allowing imaginative access to a complex reality during those short winter mornings in the Latin Quarter. Later, having become a botanist, I had to admit the existence of the two generations, a true intellectual revolution. This forced me to acknowledge that a plant is not a single organism, like animals, but two organisms: a plant with spores and a plant with gametes. For one plant to exist there must be two, the two generations, whether a tree fern, sea lettuce, lily, or redwood. One plant is always two.

One Plant, Two Generations

The two generations are not to be confused with the two sexes. In the polypody, the gametophytes can be both male and female though they usually start out as one or the other. Do male and female gametophytes look different? They are identical, indistinguishable to the naked eye. (The situation is very different in ovule-bearing plants, discussed later.) Usually, the haploid generation (gametophyte) and the diploid generation (sporophyte) look different. There are rare cases in which the gametophyte is apparently identical to the sporophyte, such as in the algae *Ulva* and *Dictyota* (Figure 56). Only the more visible of the two generations is used to define the species botanically, as in the polypody; *Polypodium vulgare* is defined by its sporophyte.

The situation is complicated by the fact that the more visible, more permanent generation is not always the diploid sporophyte. An indisputable tendency exists: The diploid generation, with its two sets of homologous chromosomes, is genetically less vulnerable than the haploid. But the opposite situation also occurs. The sporophyte of a moss is represented by each of the tiny bristles that surmount the plant by several millimeters at certain times of the year. The moss, covering tree trunks and old damp walls, is a gametophyte. What is visible and permanent, winter to summer, is the popular image of the moss, composed of branches, leaves, and tiny "roots." Numerous species, collected and cultivated with infinite care by monks, constitute one of the charming attractions of Kyōto, the botanical garden of mosses at the temple of Kokedera.

It is the two complementary generations, the gametophyte and the sporophyte, whose collaboration gives the plant species its real-

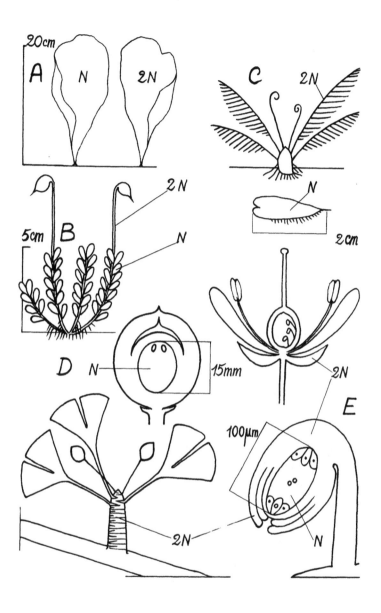

Figure 56. Gametophytes. (A) In the green alga *Ulva*, the sporophyte, 2*N*, and gametophyte, *N*, are indistinguishable to the naked eye and grow 20 cm or more long. (B) In a moss, the sporophyte is parasitic on the gametophyte, the latter growing several centimeters high and bearing foliage. (C) In a fern, the gametophyte is a tiny green blade less than 3 cm long. (D) In *Ginkgo*, the gametophyte is parasitic within the sporophyte, which is the leafy tree. The gametophyte measures no more than several millimeters long. (E) In flowering plants, the gametophyte is reduced to a microscopic embryo sac situated in the center of the ovule.

ity, in varying relationships with each other, plant to plant. Ferns and sea lettuce, horsetails and *Enteromorpha*—they practice coexistence in the complete independence of the generations, growing side by side or partitioned into slightly different environments but not directly linked to each other. Ferns and their allies have a gametophyte, the prothallus, small and ephemeral, passing its brief existence in a somber and moist little place. Its primary activity is producing gametes and, if fertilization occurs, carrying the young sporophyte. It disappears soon after having given birth to the fern that is so familiar to us.

Nevertheless, independence of the two generations is not the most frequently observed situation. In most plants, the sporophyte and gametophyte are organically linked to each other. The nature of this link is surprising: It is parasitic. That, at least, was what was affirmed by the botany professors at the Sorbonne, and as a student I was disconcerted by the apparent absurdity of a plant parasitic on itself! It was necessary for me to pull the evidence together; I tested it and the result was successful if a bit lopsided. There are two solutions to the parasitism between the generations: the sporophyte parasitic on the gametophyte or the other way around. The direction of the parasitism is not as important as the fact that the two generations remain together spatially. It is a little like the great fishes of the ocean depths in which the male, dwarfed, fixes itself as a parasite on the female, guaranteeing that the two sexes will not lose each other in the total darkness of the abyss, or more simply, like an embryo in the belly of its mother.

Mosses and liverworts—*Polytrichum, Funaria, Marchantia*, etc.—have tried out the solution of the sporophyte parasitic on the gametophyte. The thread-like stalked structure, which represents the diploid generation, is deprived of any autonomy, devoid of chlorophyll, and fed by what we call the moss, which is the haploid generation. The diploid is parasitic on the haploid moss. If it is particularly small it has undergone parasitic reduction, as in *Orobanche*, similar to the reduction in size of barnacles attached to whales. In parasites, animal or plant, the body diminishes in size in relation to its sexual organs, which become exceedingly active.

Is it legitimate to consider the moss sporophyte a parasite on the moss itself simply because it is connected to the latter during its entire life, and from which it receives all its nourishment? After all, no one would consider a wing as parasitic on the bird. However, this comparison is false for at least two reasons. The wing and the bird have the same genome whereas the moss sporophyte is genetically different from the moss carrying it. Also, the biological functions of the two generations are different. In other plant groups, evolution has made an autonomous plant from the sporophyte, green and rooted in the soil. This is true for ferns, for example. We cannot imagine an autonomous wing, however, flying on its own, separate from the bird.

Parasitic Reduction in the Haploid Generation

Cedars, firs, larches, and araucarias, otherwise known as gymnosperms, just like maize and orchids, lindens and dandelions, the angiosperms, have preferred the solution of the gametophyte parasitic on the sporophyte. Gymnosperms and angiosperms, contrary to mosses and ferns, possess ovules. These organs deserve attention for two reasons. First, they are remarkable products of plant evolution. A hollow organ produced by the sporophyte and surrounding the gametophyte, the ovule makes it clear that the gametophyte parasitizes the sporophyte (Figure 56). The central portion of the ovule, the large green prothallus of *Ginkgo* or the minuscule translucent cytoplasmic droplet of the embryo sac in the ovule of an oak or morning glory, for example, represents the gametophyte. Protected by the surrounding diploid tissue, completely sheltered from the sunlight, nourished entirely by the sporophyte, without doubt it is parasitic. Second, the word ovule deserves attention for another reason but not a very glorious one. Botanists and zoologists have never succeeded in agreeing on a single meaning for the word (Figure 57). We speak of ovulation in female animals when the female gamete becomes accessible to sperm cells, thus ripe for fertilization. As for the teaching of biology, I regret that students remember only one part of the concept of *ovule* or the other, depending in which lecture room they find themselves. I feel a bit ashamed about this lack of communication between our two biological disciplines, and it is not the only example.

Let us return to the gametophyte. Parasitic reduction is revealed; it has proven efficacious that the gametophyte is formed from genetically vulnerable haploid tissue. There has been a spectacular decline

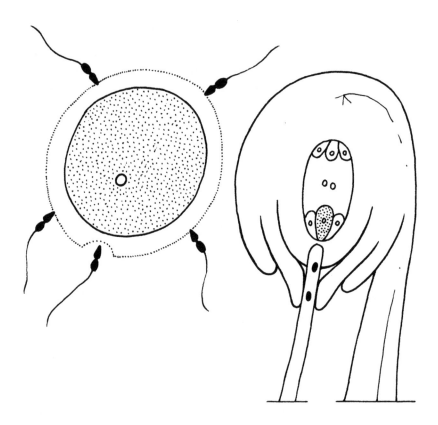

Figure 57. Left, *ovule* (French) in the zoological or medical sense is a female gamete. Right, the ovule in the botanical sense is more complex, containing the female gametophyte, the embryo sac, which itself contains the female gamete, surrounded by sporophytic tissues on which the gametophyte is parasitic. Similarities in form and dimensions stimulated use of the same term to designate the two different realities. In English, however, this confusion in terminology does not exist: *ovule* for plants, *ovum* for animals.

in the size of the female gametophyte in plants with ovules (gymnosperms and angiosperms) through the following steps: Gametophytes became parasitic in ancient gymnosperms—cycads and *Ginkgo*. They lost, along with their autonomy, all their organs—no stems, roots, or leaves—but they remained relatively large (3 cm) and green. Since they are never exposed to light, this chlorophyll may be a vestige of a time when the gametophyte was exposed to the sun. In more recent gymnosperms—pines, redwoods—the female gametophyte lost its chlorophyll and is no more than several hundred micrometers long. Finally, in angiosperms—*Populus*, *Campanula*, iroko (*Milicia excelsa*, an African rain forest tree)—the female gametophyte has lost even its cellular structure and is no more than several micrometers long (Figure 56).

This somatic reduction was not accompanied by diminution in sexual activity. On the contrary, the sexual organs, the archegonia, remained active and ended up taking most of the space with very little room left for somatic tissue. The embryo sac of flowering plants is practically reduced to a pair of archegonia. Parallel to the reduction of the female gametophyte, the male gametophyte was reduced just as spectacularly but evolutionarily much more rapidly since among the ancient gymnosperms it was already reduced to the microscopic contents of a grain of pollen.

Whatever we make of this reduction, all plants possess two genetically different generations, one haploid and the other diploid. And animals? On this point they are much simpler than plants since they have only one generation.

One Animal, One Generation

A cockroach, a tyrannosaur, a praying mantis—all are diploid organisms. Like all multicellular animals, they have only a single generation, diploid. When reproducing, they generate haploid male and female gametes in specialized sexual organs. These gametes are short-lived single cells, in no way resembling a haploid organism. These animals do not have a haploid generation, and the alternation between haploid and diploid generations makes no sense for them (Figure 55). Biology is not a refuge for uniformity, and that is part of its beauty. Laminarians, huge marine algae living just off the ocean coasts, produce gametes directly from their diploid tissue and thus have no need for a gametophyte. On the other hand, foraminiferans, single-celled animals of the marine plankton, have two types of individuals in each species. They do not represent the two sexes; they are two generations, haploid and diploid, "a rarity that up to the present is unique among animals but common in plants" (Grassé 1955).

To appreciate the extent to which plants differ from animals in this way, it suffices to say that a transposition would be as attractive as a grotesque nightmare. Imagine that to describe the life of the common swift *(Apus apus)* it would be necessary to consider, beyond the diploid swift itself, a second animal, parasitic, haploid and tiny, a sort of tick living in total dependence on the first. The swift would carry it all the way to Africa, and it would be impossible to recognize in the charming bird the great importance it owed to its ignominious parasite. One would be necessary for the existence of the other. Lunacy! Yet, remember that this is how plants are produced.

In fact, are not a caterpillar, a pupa, and an adult all required to produce a butterfly, as a prothallus and a sporophyte are to produce a polypody? Such a comparison is worthless because the same genome controls the development of the caterpillar and the butterfly, and only the latter reproduces. The caterpillar and the butterfly are not distinct individuals but stages in the development of the same individual.

Definitively, all plants require two generations, haploid and diploid, with only rare exceptions. Animals are always diploid. Thus we face a profound difference between the two kingdoms. These facts are accepted by all biologists, but an important and rarely posed question remains, on which agreement is lacking: Why do plants need two generations yet animals do very well with one? Is this a question of simple contingency, a phenomenon appearing by chance and stabilized by inertia? To me, the answer resides rather in the evolutionary mechanisms differentiating animals and plants, and I commence with the distinction between soma and germ.

Soma and Germ

More than a century ago the German biologist August Weismann (1892) made a discovery that would become the core of modern genetics and even one of the foundations of the notion of the individual. Many animals, including humans, contain two cell lines in their body plans, distinguished by their positions in the structure and by their biological functions. The somatic line, the soma, quantitatively the most significant, is responsible for the construction of the body itself, its sensory organs, its integrative systems, its limbs and viscera. When we view an animal or a human being, what we have before us is the somatic line, or soma. The germinal line, the germ, is just as important but much more discreet because it is much less abundant and not accessible to direct observation. It is, in effect, sequestered in the interior of the sexual organs, male and female. The single function of the germ line is to produce gametes.

Absolutely no exchange of cells can occur between these two lines, even if the genome is identical in the two cell types. In a leech, pig, or redbreast, the separation of these two cell lines occurs at an extremely early stage in the embryonic life of the animal. The fertilized egg is the only cell that is neither germ or soma; it is both at the same time. For example, in the roundworm *Ascaris* the two cells resulting from the first division of the egg are already different. One will give rise to the soma and the other to the germ. Even before this first division, the spherical egg cell carries somatic and germinal hemispheres— different in the physicochemical characteristics of their cytoplasm.

Ascaris is not exceptional; for most but not all animals (exceptions are mentioned in Chapter 6) embryonic development requires early migration of a series of cells that regroup in the interior of sexual organs. Sequestered thusly and becoming the germ line, they enter a period of inactivity and are activated only at the age of reproduction. Only these cells produce gametes.

Biologists at the beginning of the 20th century were strongly influenced by the fact that the two lines have contrary destinies. The soma, condemned to disappear, is only a mortal vessel whereas the germ is potentially immortal. Weismann had shown that the laws of heredity are reserved for the germ line and that the soma, doomed to death, is devoid of any evolutionary significance. His ideas swept away those of Lamarck on the inheritance of acquired characteristics, which were still influential at the end of the 19th century. Characteristics acquired from environmental influence affected only the soma, and their inheritance was thus impossible; the Lamarckian view of evolution was abandoned. The ideas of Weismann, strengthened by victory over the last Lamarckians, were applied by embryologists. They generalized about the precocious sequestration of the germ line in all vertebrates, backed up by the geneticists who discovered the universality of Mendel's laws of inheritance. This became a central dogma in biology until the mid-20th century.

In the 1950s, biologists were reminded, thanks particularly to John Harper, that their discipline is not limited just to vertebrates, or even just to animals. In reality, plants had been forgotten, beings that obstinately refused to conform to Weismann's ideas! Actually, botanists had known this for a long time. Several years after the appearance of Weismann's major work, *Das Keimplasma, eine Theorie der Vererbung,* his celebrated ideas were criticized concerning their application to plants (Buss 1983). Plants produce gametes in a manner very different from animals (Walbot and Cullis 1983, Walbot 1985).

Do Plants Have a Germ Line?

Many biologists would not hesitate to discuss the subject in this way, but I find it an exaggeration to say that plants do not have a germ line. Rather, in differentiating them from animals, we may say that plants do not have a germ line distinct from a somatic line. In other words, plants have germ and soma together, switching back and forth between the two. At first glance this seems a rather small difference. It is of extreme importance, however, and to understand why we must follow the development of a plant from the beginning.

In the egg cell of a plant—the first cell—the one resulting immediately from fertilization, there is no hemisphere specialized for the ultimate production of gametes. Nor does this occur in the plant embryo or even in the young, growing plant. In plants, there is no separate germ line. Sexual organs—sporangia, cones, flowers—are organs on the periphery, formed late by the plant and the result of meristematic production and morphogenesis. These can take a long time to appear. For many trees it is useless to look for flowers before 10 years of growth, even 20–40 years in some. Contrary to what occurs in animals, there is no precocity in the formation of cells that produce gametes. Finally, after a sometimes long sterile phase, some cells of the plant soma change into germ cells, and the plant is thus capable of expressing its sexuality. To use a nice comparison of Virginia Walbot's (1985), the situation in plants is comparable to that of an animal producing gametes from its skin!

Why do plants not have a germ line, like animals? Is this the result of chance? A functional explanation seems reasonable. In animals, gametes necessary for sexual function can be furnished by a single small group of germ cells because animals do not live very long. In a tree or a clone of a perennial plant, whose life and sexuality extend over centuries, a small group of unique cells would be incapable of providing all the necessary gametes (Walbot 1985). Over the course of the long lives of many plants, at many times and at the growing points, groups of somatic cells surrounded by sexual organs (sporangia, ovules, stamens) act like germ cells and produce gametes, more or less directly. During the entire life of a plant, the soma gives rise to numerous transitory germ lines. This is not only true of perennials; it is equally true of annuals and biennials, plants with brief lives and occurring in the minority, having appeared much later evolutionarily.

Can we find some equivalent to the true germ line of animals in plants? Can we find some analogy between the animal germ line and the gametophyte of plants? This analogy has several arguments in its favor (Vincent Savolainen, personal communication). In the germ line and the gametophyte, the tissues are highly reduced, at least in most organisms, and sequestered in organs whose function is to produce gametes. In gymnosperms and angiosperms, the sex organs are the ovule and pollen grain (Figure 57). However, the differences between the germ and gametophyte are enormous. For a gametophyte, there is no continuation to the following generation; the animal's soma and germ have the same genome and the same ploidy, whereas the haploid gametophyte is different from the diploid sporophyte.

The critical difference is that we cannot conceive of animal germ cells leading an autonomous life, whereas the gametophyte is independent, at least among more ancient plants (mosses, ferns, etc.), with normal photosynthesis and thus a completely autonomous life. For these reasons it does not seem that gametophytes and germ lines are homologous, and it would be artificial to link them. Why do plants require two successive generations whereas animals do well with only one?

The Plasticity of Organisms

In both plants and animals, plasticity represents variability in form and structure imposed by the external environment (Silander 1985, Jennings and Trewavas 1986). Which have the most plasticity, plants or animals? This question has been discussed frequently by biologists since the early 1980s, and I give here a only a sense of the discussion since each biologist contends that his or her organism holds the record for plasticity.

There are many spectacular examples of plasticity among animals, and all zoological groups have some examples to offer. The caterpillar of the butterfly *Nemoria arizonaria* feeds on oaks and produces two distinct forms, depending on the season (Figure 58). In spring, before foliage has flushed out, caterpillars feed on catkins—poor in tannins—and resemble the catkins. During summer, caterpillars feed on foliage—rich in tannins—and take on the appearance of a branch (Greene 1989). The concentration of tannin controls the change, one form or the other; that is the external factor, so here is a good example of organismal plasticity.

Among fishes, *Cichlasoma managuense* has small round teeth if fed ground food and produces large sharp teeth if it encounters living prey capable of rapid flight, such as the small crustacean *Artemia*. A change in food produces a visible modification of its teeth (Meyer 1987). Those who raise turtles know how to produce males or females by burying their eggs more or less deeply in the soil. The external factor determining sex is temperature. The shy hare of the high Alps, *Lepus timidus*, whose coat is brownish red during summer, becomes white when its habitat is covered by snow, except the tips of the ears, which remain black. In Canada, the arctic hare, *L. arcticus*, does the same. And I, poor human, know to what extent my body is plastic; 2 months without exercise and goodbye to muscles, hello to paunch! It is reversible . . . in theory.

Thousands of such examples could be given. The animal as an organism is indisputably gifted with plasticity. Is it, however, *very* plastic? It seems to me that these responses are limited, circumscribed and predictable. How do plants do?

Navigating equatorial rivers offers biologists the spectacle of the subtle struggle of trees against the glaring light, against other plants, against the current, against the submergence of banks, yet all the time growing in an environment favorable for life since neither water, warmth, nor light is lacking. Hour after hour, the pirogue moves over the brown water. Swallows drink while frolicking on the surface, the silence interrupted by the cries of birds of prey, by the plunges of iguanas falling from high branches, by the jokes of the crew in a language I do not understand. A quick downpour abates the heavy heat; the sun returns, making the nearby forest smoke. There is no lack of time to observe and try to pierce the profound mystery surrounding the nature of plants.

For most forest trees, finding their feet completely inundated would mean quick death from asphyxiation of their roots. In 15 days or so, without any apparent wounds, a great tree can thus perish.

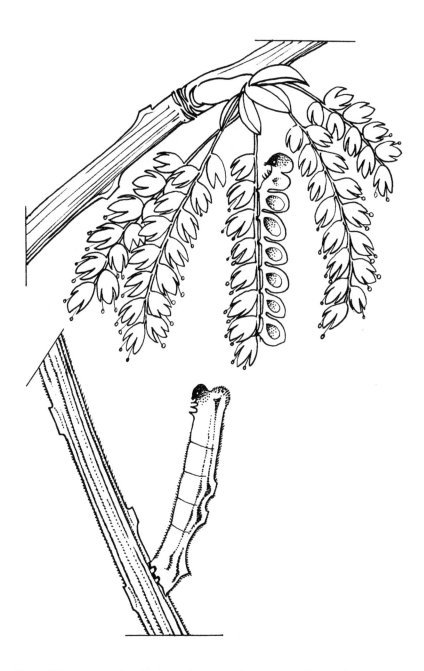

Figure 58. An example of animal plasticity. The geometrid butterfly *Nemoria arizonaria* has caterpillars that develop differently according to the time of year. Above, caterpillars appearing in spring feed on male flowers of various oak species, and they resemble male catkins. Below, caterpillars appearing during summer, after the catkins have fallen, resemble young shoots of the oaks (Greene 1989).

However, trees on riverbanks are adapted to such floods. Do they survive? They lose their leaves, slow their metabolism, waiting, months if necessary, for the water to retreat. They come out of their sleep, grow again, and put out new foliage.

We continue to move upriver. Here is a tree covered with rapidly growing lianas. This is terribly difficult for the tree (Putz 1994), which is in danger of dying for lack of light this time. Here also, patience pays. Could the liana be killed by the flood, which thus becomes the ally of the tree? Could it be that the liana grows quickly, then dies quickly? Could it be that the tree is able to pierce the mantle that suffocates it and deprives it of light, producing a vigorous shoot that will save its life? The reader will remember that reiteration (Chapter 2), a property possessed by trees, not animals, allows their developmental plan to be repeated several times (actually without limit). To give a specific example, a tree growing on a riverbank is undermined by the current and topples. It finds itself horizontal, its branches under water. If the bank is a small cliff 20–30 m high, as it often is along tropical rivers, the tree could reestablish its crown (Figure 59). A tree, having lost its natural ascending vertical orientation, could die if its roots are damaged or its foliage is immersed in the river. If the roots and tip are not too damaged, the tree regains its balance through reiteration. If these shoots, called reiterations, grow freely long enough, they will form new small trees, derived from the parent, replacing it.

Duhamel de Monceau (1758), turning a tree upside down, noted the growth of roots on submerged branches and leafy branches on roots sticking up in the air (Figure 59). The English botanist Paul Richards (1952), exploring forests in Nigeria, remarked on the strange alignments of trees, all of the same species and in rows. Was this the result of jesting aboriginals who planted them in the primary rain forest, lines of trees resembling those bordering the streets of Kent? Richards understood that it was a natural phenomenon, the reiteration of trees knocked down by wind. Larson, Kelly, and Matthes-Sears (1995), biologists working in Canada, showed the incredible structural differences between trees growing on horizontal soils and those of the same species growing in crevices on a vertical cliff (Figure 60).

There is no need to travel great distance to observe plasticity in plants; the common dandelion holds a sort of record for this. It can

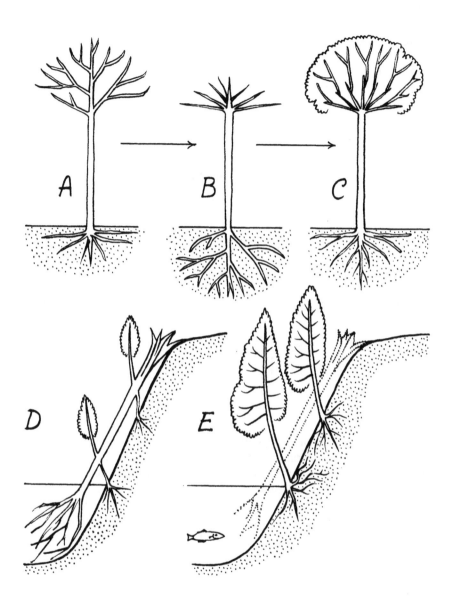

Figure 59. Examples of plant plasticity. (A) A vigorous young tree in winter dormancy is turned upside down so that its roots are in the air and its branches in the soil (B). (C) During the following period of growth, the tree reestablishes its normal silhouette (Duhamel du Monceau 1758). (D, E) A tree fallen into a river can produce reiterations that will replace it before it decomposes.

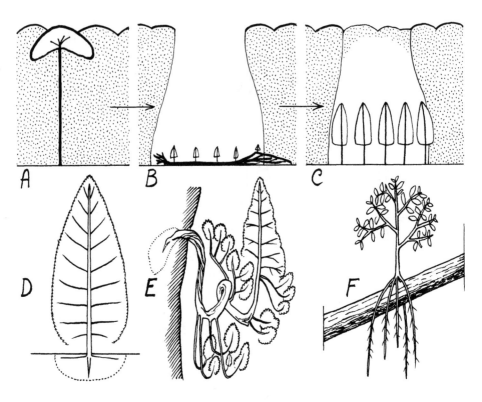

Figure 60. Further examples of plant plasticity. (A) In a tropical forest, a great tree is felled during a storm. (B) At the bottom of the opening caused by the fallen tree, the trunk produces reiterations along its length. (C) Some years later, the fallen trunk has disappeared and a line of identical and contemporaneous young trees persists (Richards 1952). (D) A tree grows in favorable conditions. (E) A tree of the same species grows in a crevice in a vertical cliff (Larson et al. 1995). (F) In a tropical forest, a reiteration appears on a leaning trunk, often producing a clearly visible system of lateral roots.

grow and flower almost anytime, even during the weak sunshine of winter. Despite nearly incessant growth, it never grows above the soil surface, its roots pulling it down. By this strange negative growth it avoids an erect form. Clinging to the interface between earth and sky allows it to benefit from the more moderate conditions there than those experienced by erect plants. Thus it can invade all the continents, survive all climates except the hottest; the dandelion has become truly cosmopolitan.

Plant combativeness is revealed when we try to destroy them. What do we have to do to get rid of dandelions? Cut them right down to the soil? They grow again. Tear them out? Their fine roots regenerate plants. Bury them under a thick layer of soil? A long shoot leads up to the surface at the same place. Reduce them to tiny pieces? That is a bad idea. Each piece of stem or root will regenerate a dandelion, and where there was only one, it comes back with some buddies.

These examples give an idea of the range of plasticity of plants and its adaptational value. Placed in clearly unfavorable conditions, plants die, but if survival is possible they overcome difficulties with an obstinacy that compels our admiration. If I would dare use an animal metaphor, I would speak of their courage. They do not fight to survive but to retake their place in the vegetation. If conditions again become favorable, they succeed in erasing any trace of difficult times, aided by their autonomy, by the fact that they experience no difficulty in changing their form, and by the time at their disposal.

Who Wins the Prize for Plasticity?

It seems to me that the range of plasticity in animals is limited compared to that in plants. In its animal version, plasticity is manifested as a sort of prolonging of development. Favorable external conditions cause certain genes to be expressed, and if conditions become unfavorable, other genes are expressed and the organism adjusts its functioning to the new conditions.

Plant plasticity is an entirely different matter. There are no limits to the type of adjustment, which often attains a degree of true upheaval. Take the tree of Duhamel du Monceau, surviving in spite of its buried branches and exposed roots. Then find an animal under such extreme conditions, feeding through its anus and defecating through its mouth! Through such comparisons, though sometimes difficult because the situations may not be comparable, a consensus may be established: An organism is more plastic if it is a plant than if it is an animal. We should suspect the reason, which has already been discussed: Plasticity in plants can substitute for mobility (Grime et al. 1986).

Confronted with predators, tempted by prey, or troubled by ecological change, animals find the solution to their problems through mobility: fleeing to evade pursuers, hunting to assuage hunger, seeking a better environment. We understand that moving animals have no need of developing great plasticity precisely because they are mobile. On the contrary, plants must adapt in place to survive, and plasticity provides adaptive responses to variations in the environment. More fundamentally, without doubt, plasticity also occurs in the genome. The situation in animals is very different from that in plants.

Genomic Plasticity

For free-living, mobile animals, including humans, genetic homogeneity and stability are important to the individual (Slatkin 1987). From our ears to our prostate, from epiglottis to neurons, almost all our cells have the same chromosomes, the same diploid state, the same distribution of genes, in brief, the same genome. That is true at this instant and remains so over the course of time. My cells as an infant had the same genome as my cells as an adult, or the cells of the old man I will become.

In biology, generalizations are imprudent, so it is advisable to mention examples such as benign somatic mutations, transposable elements (jumping genes), the giant chromosomes of flies, haploid males in hymenopterans, and polyploid hepatic and immune surveillance cells. It is important to note that these are very special cases, so it remains true that homogeneity and genetic stability of the individual—in space and time—have value as dogma in zoology and medicine. For animals, including humans, this genetic homogeneity and stability are particularly significant; they bear on the search for the fundamental notion of the individual.

Animals are usually intolerant of variation in their genomes. In humans, serious genetic or chromosomal anomalies are responsible for 50% of spontaneous abortions. Three copies of chromosome 21 instead of two produces Down's syndrome. The expression of a mutation that makes a cell malignant can, in time, lead to death. Certain genes undergo mutations so that affected individuals suffer from afflictions as serious as albinism, hemophilia, achondroplastic dwarfism, color blindness, congenital subluxation of the hip, myopathy, cystic fibrosis, congenital blindness, or Huntington disease. Benign mutations also occur, such as those that produce birthmarks or nevus, but these are in the minority. I cite human examples because our genetics, of direct interest to us, is relatively well known. It can be generalized to other animals: Animals tolerate incidental modifications to their genomes poorly if at all. Their genomes are deficient in plasticity.

There is other evidence of the absence of genomic plasticity in animals. For example, it is nearly impossible to cross different animal species, even closely related ones, and obtain a viable and fertile descendent. The immune system, which removes all nonconforming cells, provides evidence for intolerance of genetic variation in animals, no matter where it originates.

Molecular genetics is replete with sophisticated mechanisms for DNA replication in animals. I have been particularly impressed by the $T53$ gene, which controls production of P53 protein, blocking the mitosis of cells whose DNA is abnormal. From the hopes it raised in the war on cancer, P53 garnered the nickname "guardian of the genome" and was named Molecule of the Year by *Science* magazine in 1993.

Is there a relationship between the mode of life in animals and their lack of genomic plasticity? Does one explain the other? Why would mobile animals need plastic genomes? An animal whose environment, in the broadest sense, changes in an unfavorable way often finds the solution through reversible changes in its physiology or behavior. When day arrives, bats sleep under roof tiles; in winter, the hedgehog hibernates; during the dry season, the lungfish hides in the mud of the shallows and slows its metabolism. Animals resort to mobility even more frequently to maintain an adequate environment. At the end of summer, swifts fly from temperate regions to live in the Tropics; in the dry season, lions in East Africa gather around water holes; pursued by a tuna, the flying fish leaps out of the water and glides several dozen meters. Whether a question of fleeing or of modifying its physiology, there is nothing that requires an animal to change its nature—its identity—nothing that would require genetic change. In animals, the lack or limitation of genomic plasticity is compensated for by ample behavioral plasticity, which allows animals to resist change in their environment, giving them greater freedom from ecological constraints.

And plants? Their genomes are very different from those of animals, including humans. Genome size in plants varies dramatically from one species to another. In flowering plants, the size of the genome varies considerably, from 0.2 picogram of DNA in *Arabidopsis thaliana* to 127.4 picograms in *Fritillaria assyriaca* (Bennett et al. 1982). Mammals all have genomes weighing nearly the same, 2.1 picograms, which is even more odd considering that mammals appeared some 300 million years ago whereas flowering plants have a history of just 150 million years (Walbot and Cullis 1985). Considering that plants are 50 times more numerous than mammals, we can suppose that genomic evolution has been more rapid among them, favoring less genetic stability.

The first evidence for plasticity in plant genomes was the work of Barbara McClintock on maize *(Zea mays)* in the 1940s (Chapter 1). After more than half a century, we have an improved overview, though still incomplete, of plasticity at the chromosomal level as well as at the level of genes and their expression. Plants are much more tolerant of chromosomal modification than animals. In cultivation, haploid maize plants occur from time to time. They are distinguished only by their puny size; nothing else of importance distinguishes them from the normal diploid maize plants surrounding them (Klekowski 1988).

In ivy *(Hedera helix)* a plant is diploid when young, tetraploid when adult (Figure 61; Kessler and Reches 1977). Generally, changes in the number of sets of chromosomes do not influence performance. Haploid, diploid, and tetraploid plants are viable and fertile, or they can become fertile after regularization of chromosome number. The same is true for trisomic or monosomic plants (Walbot and Cullis 1983) though these same anomalies are almost always fatal in animals (trisomy for chromosome 21 leading to Down's syndrome and sterility in humans, for example).

At the level of the gene, plants and animals are sharply distinguished by differences in sensitivity to somatic mutation. Animals do not tolerate such mutations whereas plants generally tolerate them very well because they know how to use them. Even better, plants have clever mechanisms for increasing the frequency of somatic mutations, favoring the likelihood of their survival when their effects are favorable or neutral yet assuring their elimination when they are unfavorable. Let us first examine the mechanisms responsible for establishing mutations; sorting out of mutations follows.

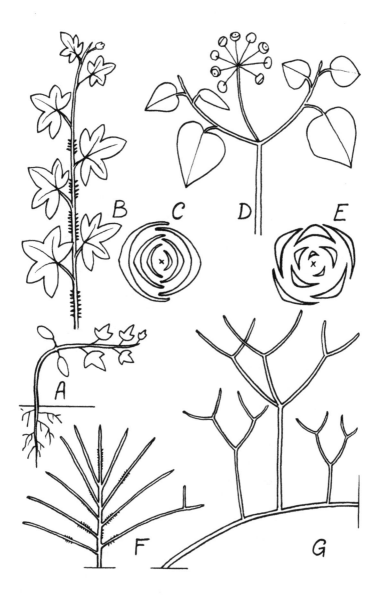

Figure 61. The two forms of ivy, *Hedera helix* (Araliaceae). (A) A young plant, which rapidly adopts a running orientation. (B) The juvenile form with lobed leaves, attached to its support by adventitious roots. (C) Distichous phyllotaxis of the juvenile form, as seen in cross section of the terminal bud. (D) The adult form, erect and reproductive, with simpler leaves. (E) Spiral phyllotaxis of the adult form. (F) Architecture adopted by a juvenile when it becomes attached to a vertical support. (G) Part of a horizontal axis in which the characteristic architecture of the adult ivy is shown. The juvenile form is diploid, and the adult form tetraploid (Kessler and Reches 1977).

Generators of Genetic Diversity

Walbot and Cullis (1983) described the mechanisms through which plants favor somatic mutations as "generators of diversity." Meristematic growth, because it is characteristic of plants and conditions their mode of life, should be considered first. The meristems, groups of cells that permanently maintain an embryonic nature, have a well-known role in the formation of tissues that make up the adult plant and in the establishment of plant form through morphogenesis. However, meristems are less well known for producing mutant cells, which a plant may require to cope with the vicissitudes of its environment.

Several authors have attracted attention to meristematic action as a source of mutant cells (Klekowski and Kazarinova-Fukshansky 1984a, b, Walbot and Cullis 1985, Gill 1986). There are two kinds of meristems: those of algae, and ferns and their allies, which have one large, easily identifiable apical cell, and those of seed plants, more evolutionarily advanced, containing numerous embryonic cells (Figure 62). All cellular division is an opportunity for mutation through error in copying during replication of the genetic information. Since meristems are where cell divisions occur, they are the sites of production of mutant cells.

Recall the immobility of the vegetable cell; plants are thus sheltered from metastases, ensuring that only a cell dies if a mutation is lethal, not the entire plant (Klekowski 1988). Contrary to animals, plants live with most of their mutations. After the appearance of a mutant cell, the events that follow depend on the size of the meristem.

In the smallest meristems (algae, ferns, etc.) a mutation of the single apical cell is passed on to all cells descending from this initial cell, which means that the entire summit of the axis is invaded by the mutation. In meristems of medium size, formed from a small group of embryonic cells, mutations are established more rapidly than in large meristems (Klekowski and Kazarinova-Fukshansky 1984a, b). In large meristems composed of several dozen embryonic cells, if a single mutation occurs, its descendants form a mutant line. The latter is surrounded by numerous other, normal cell lines, and the meristem becomes a genetic chimera. Even though its usage is well established, the zoocentric term chimera does not please me. A chimera is a monster mentioned by Homer in the *Iliad*, with the

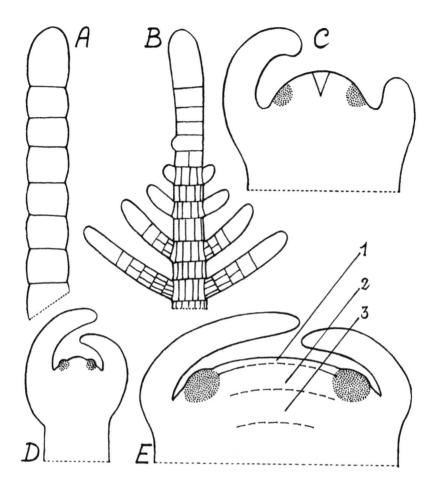

Figure 62. Forms and sizes of meristems. (A) The alga *Ulothrix* grows in a diffuse manner by division of any of its cells. (B) The alga *Sphacelaria* has an apical cell responsible for its strictly apical growth (Ducreux and Le Guyader 1995). (C) The center of the meristem of the fern *Pteris cretica* is occupied by a large apical cell of inverted pyramidal form (Michaux-Ferrière 1974). During growth of a seed plant, angiosperm or gymnosperm, from embryo (D) to young plant (E), the meristem can multiply its surface area by a factor of 20 (Klekowski and Kazarinova-Fukshansky 1984b). (E) In a meristem composed of three strata, only the middle stratum produces gametes, and only a mutation in that stratum could thus be passed to descendents (Gill 1986). The gray zones in C–E are regions of maximal mitotic activity, where leaves are produced.

head of a lion, the body of a goat, the tail of a serpent, and breathing fire. It has become a symbol of incoherence, of false ideas, of empty imagination (Borges and Guerrero 1957). The reader will note that when they do not follow the rules of zoology or conventional logic, plants are seen as incoherent and baptized with a term evoking absurdity, even monstrosity.

Depending on the nature of the mutation—its effect on the cell, particularly on the frequency of cell division—competition takes place within the meristem between the mutant and other cells (Klekowski 1988). If the mutation is unfavorable, the mutated, less active cells become a lineage without a future and the mutation is lost (Figure 63). Or it may be that the mutation favors cell division, and the mutated cells proliferate more quickly than others; then the entire meristem becomes the mutant. This hegemony and its opposite (complete loss of the mutation) rarely occur because most mutations do not affect cells, particularly their division, so directly. The mutant cells typically divide as frequently as the nonmutants, leading to the appearance of stable genetic chimeras (Figure 63). The existence of such chimeras has consequences well known to horticulturists. The same plant is capable of supporting, side by side, both normal and mutant branches. It may be that the mutant branch is, for some reason or other, more interesting to the horticulturist. It can be isolated, multiplied by rooting or grafting, and a new cultivar is born. Thousands of such mutations have thus been selected and conserved by humans for nutritional, commercial, or aesthetic reasons. Most cultivated plants have such an origin, for example, high-yielding potatoes, red grapefruit, and dwarf bananas (Whitham and Slobodchikoff 1981). This gives an idea of the frequency of such mutations in plants, but the true frequency is greater because most mutations pass unnoticed since they do not affect form, color, or yield. Also, some mutations are unfavorable, and plants eliminate them. There are other issues to discuss later, but I will not anticipate them here. Now, I wish to mention four other mechanisms that generate genetic diversity: mutation frequency, longevity, the mutagenic effects of ultraviolet radiation, and other effects caused by a diversity of constraints to which plants are subjected.

Given that we know little about the average frequency of mutation, it is interesting to note that in potatoes *(Solanum tuberosum)* and maize *(Zea mays)* the frequency is 10^{-4} and 10^{-5}, respectively,

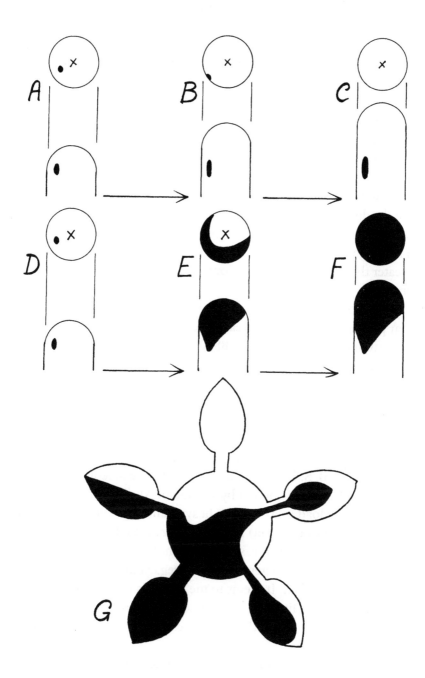

Figure 63. Competition at the level of the meristem between mutant (black) and nonmutant (white) cells. (A–C) Elimination of a mutation. (D–F) Invasion of a meristem by mutant cells (A–F, Klekowski 1988). (G) A mutation persists without taking over the meristem; a stable genetic chimera is established. This would be the most probable outcome (Whitham and Slobodchikoff 1981).

and in the fruit fly *(Drosophila)*, 10^{-5} to 10^{-7} (Whitham and Slobod-chikoff 1981). Generalization is not possible; we can simply state that a higher level of mutation in plants versus animals would not contradict the skimpy data available.

Longevity differentiates plants and animals. With the exception of a few specialized forms—annuals and biennials—plants do not have a determinate life span and many are potentially immortal. Animals, at the maximum, live some dozens of years, whereas trees thousands of years old are not exceptional. What is important is the not the duration of life itself but the augmentation of genetic diversity through longevity, as explained through meristematic function. The longer a plant lives, the longer its meristems are active and the greater the number of cell divisions separating the egg from the end branches.

Trees, organisms of great longevity, given the number of years that pass as their height increases, accumulate relatively more somatic mutations (Klekowski 1988, Charlesworth 1989). The meristems at the top of a great tree are thus richer in genetic chimeras. These mutant alleles, if more and more frequent in a genome, risk lowering the fitness of the organism. Should we speak of genetic load, as Klekowski (1988) has? Genetic load is a measure of the selection exercised on a genetically variable population. This definition, from Crow and Kimura (1970), works better for animals than plants. J. B. S. Haldane, the author of the concept, was a zoologist; the ideas were applied by Hermann Muller to humans, for whom genetic load is evidently a heavy handicap. Isabelle Olivieri (personal communication) has defined genetic load as the price each generation pays for superior genotypes, which are then selected. By speaking of genetic load in regard to plants, we are giving in to zoocentrism. It seems to me that it is rather a question of a reserve of variability that plants need to survive in a fluctuating environment from which they cannot escape, as discussed in more detail later.

We know that ultraviolet radiation increases mutation rates. The aerial positions of apical meristems, situated on the most exposed parts of the plant, would predispose them to be affected by ultraviolet light. Could this architecture help plants increase their genetic variability? We can say that animals protect their embryos whereas plants expose their meristems to mutagenic radiation. Canopies of tropical rain forests should be particularly affected. Ultraviolet radiation is particularly intense at tropical latitudes, precisely where the flora is most diversified.

However, ultraviolet radiation is only one example of the many constraints affecting plants, given that they live fixed in place, submitting to risk. These constraints are varied (McClintock 1984) and include nutritional deficiencies, insect penetration (leading to gall formation), viral infection, poisoning by various toxic chemicals, or even the stress caused by combining dissimilar genomes in a plant through hybridization or tissue culture.

Tissue Culture

Animal tissue or cell culture gives birth to stable lines that can be exchanged between laboratories. These lines can be cultured indefinitely yet keep their characteristics. Plants behave differently. It is well known among specialists that placing plant tissues or cells into culture has a profoundly destabilizing effect on them, leading to the appearance of many variants, often difficult to interpret, but many of which are heritable. Toward the end of the 1950s, with the commercialization of plant tissue culture, the abundance of these variations was seen as a handicap to cloning and propagating good varieties. Later, it was enthusiastically viewed as a source of new and useful varieties (Meins 1983, McClintock 1984, Gill 1986).

In the context of genetics, what is important is the lack of stability in plant tissue culture. "The few cases studied in detail show that heritable changes arising in culture can affect qualitative traits as well as quantitative traits and can result from epigenetic changes, mutations in specific gene loci, chromosomal rearrangements, and less well defined gametic modifications" (Meins 1983).

Hybrids Between Species

Animals are not very talented in hybridizing interspecifically because such crosses are usually not viable. Of course, there are some exceptions. Some hybrids have been obtained by crossing different animal species, including trout × salmon, donkey × stallion = hinny, donkey × mare = mule, dromedary × llama, and lion × tigress. With the exception of fishes, interspecific animal crosses are not easily accomplished or common. None occurs naturally, even in captivity, and it is necessary to use a variety of tricks because animals of these different species "never seek out each other spontaneously" (Bertin 1950). In all cases, including fishes, interspecific animal hybrids are not very fecund; they are barely fertile, even sterile. Their survival in a natural environment would be improbable.

Plants differ profoundly from animals in regard to hybridization. Crosses are frequently possible between plants of two distinct species. Palms may cross (Figure 64). Plane trees in the cities of southern France are the result of hybridization between two species with distinctly different geographic origins. One *(Platanus orientalis)* is a small tree that grows near rivers and other bodies of water, from Greece to Afghanistan. The other *(P. occidentalis)* grows along the Mississippi and is the largest tree of eastern north America. When the eastern and western planes found themselves together in an English botanical garden, probably Oxford, their spontaneous hybridization gave birth to the London plane *(P. ×acerifolia)*, an interspecific and fertile hybrid that became the maple-leaved plane of the Midi (André Vigouroux, personal communication). This tree benefits from a hybrid vigor that allows it to tolerate dry conditions, too much watering, urban pollution, and a disease fatal to its parents, anthracnose. Hybrid vigor helps the plane resist the chainsaw gluttony of tree surgeons, swallow up road signs, and survive even if watered with motor oil!

Even if hybrid vigor is usually not as spectacular as that of the hybrid plane, it is not a rare phenomenon in plants. Crosses may be immediately fertile, as in the case of the plane, or they can be made fertile by polyploidy (Figure 86, Chapter 6). Either way, abundant offspring may be produced. Agronomists and horticulturists have also been able to produce hybrids between different plant genera experimentally, for example in conifers, *Cupressus × Chamaecyparis* = *×Cuppressocyparis*, cereal grains, *Triticum × Secale* = *×Triticale*, or orchids, *Laelia × Cattleya* = *×Laeliocattleya*. One could object that orchids have been divided into too many, too closely related genera, explaining the ease of intergeneric crossing in the family. There may be truth in that, but not for our example. *Cattleya* has been crossed not only with *Laelia* but also with *Brassavola, Broughtonia, Diacrum,* and *Epidendrum;* it is not possible that these five genera could be confused with *Cattleya.*

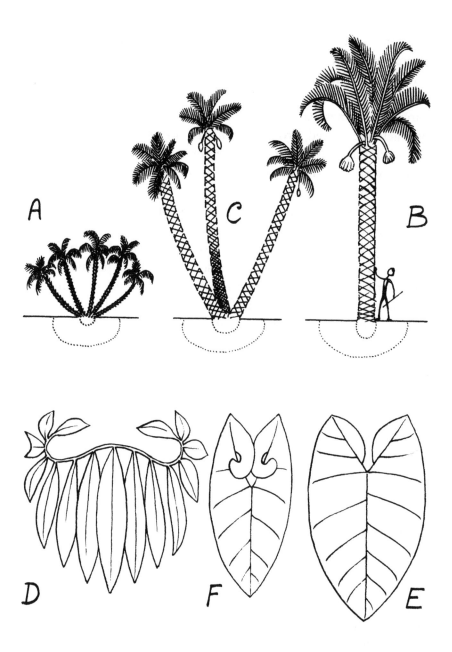

Figure 64. Interspecific hybridization in plants. (A) Senegal date palm, *Phoenix reclinata*. (B) Canary Island date palm, *Phoenix canariensis*. (C) The hybrid, *Phoenix reclinata* × *P. canariensis* (Sarasota, Florida, 1994). (D) Leaf of *Philodendron goeldii*. (E) Leaf of *Philodendron solimoesense*. (F) Leaf of the hybrid, *Philodendron goeldii* × *P. solimoesense* (Joep Moonen, personal communication). The philodendrons are from French Guiana.

Interspecific and intergeneric crosses may produce stressful situations that favor genetic diversification. Chromosomes undergo translocations, inversions, losses; they can fragment or, contrarily, fuse into fewer but much larger megachromosomes. One of the parental genomes may be partially or totally eliminated, or the hybrid genome altered, producing tumors. So strong is the effect of stress from the combination of different genomes that, "Undoubtedly, new species can arise quite suddenly as the aftermath of accidental hybridization between two species belonging to different genera" (McClintock 1984). We are confronted by a formidable mechanism for generating genetic diversity.

Finally, given that it is not proper to speak of a mechanism without mentioning occurrences of its absence, recall that plants do not have a true immune system. Contrary to what occurs in animals, from insects through vertebrates, no mechanism aimed at the elimination of cells with genomic modifications is found in plants; this favors the maintenance of genetic diversity within the individual plant. After having reviewed the generators of diversity (Walbot and Cullis 1983), we need to examine their effect, in other words, measure the genetic differences within a single plant. Sadly, this research has rarely been done directly. The results at our disposal suggest that different organs in the same plant can maintain important differences in their genomes.

Genetic Diversity Within the Plant

Differences can occur at the level of the chromosome, where they are sometimes of astonishing magnitude. In the North American wildflower *Claytonia virginica*, spring beauty, the number of chromosomes varies from 12 to nearly 200 within a single plant (Lewis 1970a, b). Intraplant genetic differences can also occur at the level of the gene. The little spearwort, *Ranunculus flammula*, is a buttercup common in wet locales in Europe; its aerial leaves are entire but its submerged leaves are reduced to veins (Figure 65). The gene controlling the development of the blade occurs as two distinct alleles, one expressed in the submerged portions of the plant, the other in the aerial portions (Cook 1974).

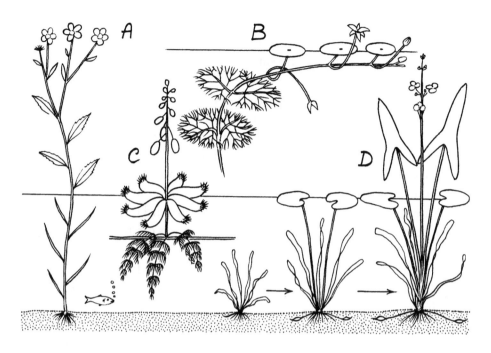

Figure 65. Genetic differences within aquatic plants. (A) In the little spearwort, *Ranunculus flammula* (Ranunculaceae), the gene controlling leaf development is found as two different alleles, one expressed in submerged portions and producing filiform leaves, the other in aerial portions and producing large leaves (Cook 1974). The little spearwort, with its two forms of foliage, represents a common situation in aquatic plants. Comparable development is found in *Cabomba* (B, Cabombaceae) and *Utricularia* (C, Utriculariaceae). (D) Arrowhead, *Sagittaria sagittifolia* (Alismataceae), has three kinds of leaves: ribbon-like submerged aquatic leaves, floating heart-shaped leaves, and arrowhead-shaped aerial leaves (Aline Raynal, personal communication).

The Strangler Figs of Lake Gatún

Practical difficulties in documenting such differences mean that measurement of intraplant genetic diversity is rarely performed on trees. Because of this, the research of Thompson et al. (1991) is particularly important and I analyze their results in detail. Their research concerned strangler figs, certain species of the genus *Ficus*. These trees are among the strangest of tropical plants. A bird deposits some seeds of the strangler fig through its droppings into the fork of a tree. The seed germinates, and as the young fig grows leafy shoots up, it sends roots to the ground. The roots increase in thickness, fusing with one another and forming a ligneous network that surrounds the trunk of the supporting tree. The latter, unable to grow, dies and eventually disappears. The status of the fig thus passes from a strangling hemiepiphyte to an erect tree, supported by a strange, reticulated, hollow trunk, forming a memory of its earlier support long after the latter has disappeared (Figure 66). Thompson and collaborators studied genetic variability in strangler figs on Barro Colorado Island in Lake Gatún, Panama. They chose the largest trees, whose numerous branches were accessible by boat. Since these trees grew on banks and leaned toward the lake surface, we can deduce that this genetic investigation was carried out, in fact, in the canopy.

Thompson and coworkers selected young leaves from each of the large branches of a fig, dried them, and kept them cold. Ultimately, these leaves were analyzed by electrophoresis, which allowed testing of the variability of genes in 18 enzyme systems. They tested 14 trees for genetic differences between branches, belonging to six species: *Ficus citrifolia*, *F. colubrinii*, *F. costaricana*, *F. obtusifolia*, *F. perforata*, and *F.* cf. *trigonata*. The results were spectacular. With the exception of one individual of *F. citrifolia*, the 13 other trees were mosaics of several genotypes. All six species varied in the alleles present in branches of the same tree. Collectively, these 13 variable fig trees contained at least 45 different genotypes—45 genotypes for 13 trees!

The interpretation by Thompson and coworkers was as surprising to me as their result, that these large trees came from the fusion of several small trees. The hypothesis of somatic mutation was discussed briefly, and it was admitted that mutations would be consistent with the observed genetic diversity but that the mutation explanation appeared too broad—as many as four loci were implicated—for that hypothesis to suffice. Knowing the rapidity with which roots of strangler figs fuse together, they preferred to explain the diversity by fusion of separate treelets. Seeking to verify their interpretation, they traced anatomical connections between the roots and shoots of one tree using colored liquids. They found that each root connected to a single branch, which they interpreted as demonstrating little physiological integration for the composite tree. They concluded that these results supported the fusion hypothesis.

Figure 66. A strangler fig, *Ficus nymphaeifolia* of French Guiana. Left, having germinated 18 m up in the fork of a supporting tree, the young fig sends a taproot toward the ground. Center, the crown of the fig is larger than that of the support tree, which is beginning to die. The root network of the fig surrounds the trunk, impeding growth in diameter. Right, the support tree is dead, and the fig itself has become a huge tree with a networked trunk shown in cross section (Juliana Prosperi, personal communication). The surface roots of the fig, indicated by the horizontal arrows, are 130 m long (Claire Atger, personal communication).

This investigation of genetic variability is fascinating and extremely novel, but I do not think that the fusion hypothesis is really confirmed. Certainly, it is not very prudent of me to have an opinion about trees I have never seen. Even if I am not familiar with the figs of Lake Gatún, I have observed strangler figs in many regions of the world, and the architecture of tropical trees is one of my scientific specialties. On that basis, I take the liberty of formulating some criticisms of the interpretation by Thompson et al. (1991) of the trees with several genotypes.

Roots of strangler figs are notorious for fusing quickly, but these fusions occur between roots of the same fig rather than between those of different individuals. Are fusions between trees of the same species, in technical terms an allograft, possible in figs? Some have been observed (Putz and Holbrook 1986) but are not considered common. Personally, I have never observed it, but allografts seem possible in *Ficus* as well as in other tropical trees. But allografts in *Ficus* must be rare, even exceptional, based on what I know of tropical trees in general. In my opinion, it is unreasonable to conclude that six species (and 13 trees) in one small area would fuse in such a way.

As for weak physiological integration between subsidiary branches in the tree, this is true and been shown in several studies (Shigo 1991). It is not proof of allografts but is a consequence of the colonial nature of reiterated trees (Chapter 2).

The superb results of genetic diversity within a tree obtained by Thompson and colleagues appear to me to be explained better by somatic mutation rather than allografts. In my opinion, with reservations until there can be a complete evaluation of the details of the problem, the subjects connect in the following way: Animals, including humans, having uniform genomes, and the human or animal orientation being dominant in biology, these investigators have assumed that a tree must also have such a genome. If a huge tree is revealed to have several genotypes, the conclusion is that it is the result of precocious fusion of several small trees. Certainly, such a Procrustean attitude discounts the fluidity of plant genomes, revealed by Barbara McClintock and further documented by many others whose research has contributed to knowledge of the phenomenon. The interpretation by Thompson et al. (1991) remains for me an example of the handicap that the zoological perspective exercises on our understanding of plants.

[There is a problem with this interpretation. The amount of genetic diversity sampled by the 14 enzyme stains is extremely small compared to the total number of alleles, and the probability is exceedingly remote that such a survey would detect any differences between the branches (Antolin and Strobeck 1985). Although these data are not inconsistent with the existence of genetic variation within tree crowns (James Hamrick, personal communication), much finer techniques would be necessary to reveal it. —translator's comment]

In 1996, during an expedition to explore the rain forest canopy in French Guiana, Darlyne Murawski attempted to measure genetic diversity in the crowns of individual trees directly. The trees selected were two huge legumes, a timbauba *(Enterolobium schomburgkii)* and a yellow Saint-Martin *(Hymenolobium flavum)*. The latter, more than 50 m high, is certainly one of the great trees of the coastal forests of the Guianas and is the focus of discussion here. The canopy raft carried by a hot-air dirigible allowed us to move around the crown of the old giant to collect leafy shoots at the tips of the major branches (Figure 67). Under the fuselage of the dirigible, Darlyne mapped the enormous crown to establish the location of each collection. Then the analysis of the tree's genome was performed at the Gray Herbarium of Harvard University.

DNA was extracted from the specimens and purified. Fragments of DNA were produced and tagged for identification, then amplified using the polymerase chain reaction. Finally, the amplified fragments were separated by electrophoresis and made visible under ultraviolet light. This technique (randomly amplified polymorphic DNA) is sensitive enough to detect differences in genomes between branches of the same tree, and the result was clear. The *Hymenolobium flavum* contained several clearly different genomes. They were not randomly distributed within the tree crown but were grouped in sectors. *Enterolobium schomburgkii* produced an identical result. Darlyne and I believe that we have established the existence of genetic variability within the crown of a tree (Murawski 1998) but many more tests will be necessary to map the distribution of genomes in tree crowns precisely. This confirms the concept that Slatkin (1987) advocated for plants, that they are organisms that do not sequester germ cells at an early stage of development, instead producing numerous reproductive organs with the possibility of genetic variation in different parts of the same individual as a result of accumulation of somatic mutations during the course of development.

If we summarize the mechanisms that generate genetic diversity in plants—indeterminate meristematic growth, higher levels of mutation than in animals, longer life spans, the mutagenic activity of ultraviolet radiation on exposed plant organs, other stresses—it is realistic to ask whether the initial genome of the egg cell persists till the end of life of a large tree. I believe that no scientist, at present, can document an answer to this difficult question. I add just two remarks. One, plasticity may be an inadequate term for describing the plant genome; Walbot and Cullis (1985) suggested the term fluidity. Two, even if it is likely that a tree's genome changes during its

Figure 67. Genetic heterogeneity within a tree. Within the crown of a large leguminous tree, *Hymenolobium flavum,* a raft suspended beneath a hot-air dirigible is used to collect samples destined for genomic analysis (operation of Radeau des Cimes, Paracou, French Guiana, November 1996; Murawski 1998).

lifetime, the tree certainly remains itself, and that is an important difference between plants and animals.

How does a plant remain the same over time as its genome changes and diversifies ceaselessly? It seems useful to approach the subject as a hypothesis, one that is gaining ground among biologists (Haber et al. 1961, Sachs 1978, 1988, Hagemann 1982, Lintilhac 1984, Kaplan and Hagemann 1991, Kaplan 1992), that plant form can be established without the control of genes. Genes would control the biochemical constitution of the plant, but not its morphology. The development of the plant would thus be an autocatalytic process (Sachs 1978), fundamentally different than that of an animal (Lintilhac 1984). The very early embryonic establishment of an element specific for managing growth and form—a genetic determinism—is prolonged through development, taking all that has preceded as a guide. This initial managing element would thus become the unique morphogenetic reference, and growth would not call on genes *at this level*. For Lintilhac (1984), "Shape begets shape." All this would occur if the form of a plant carried information sufficient to substitute for the genome. This is food for thought. It is not yet possible to draw a complete picture from the disparate elements, and the subject merits further study.

If plants only had mechanisms to generate genetic diversity, their reproduction would cease to be true to the original. But it is easily seen that this is not true: An oak bears acorns from which more oaks grow. Plants also have mechanisms to limit genomic plasticity.

Sorting Mechanisms

An example of discrimination against a cell line concerns mutations that limit the synthesis of chlorophyll. To the degree that the mutant cells are not gravely handicapped, thanks to nutrition furnished by neighboring cells that have normal photosynthesis, an easily observed genetic chimera is obtainable. This is a phenomenon well known to horticulturists because of its decorative qualities. Such plants, in which normal green photosynthetic tissue is juxtaposed with white tissue whose cells lack chloroplasts, are variegated (Figure 68A, B). Everyone has seen variegated plants, sold in supermarkets along with other ornamental plants: *Ficus benjamina, Sansevieria*, ivy, periwinkle, *Dieffenbachia*, Japanese spindle tree, etc.

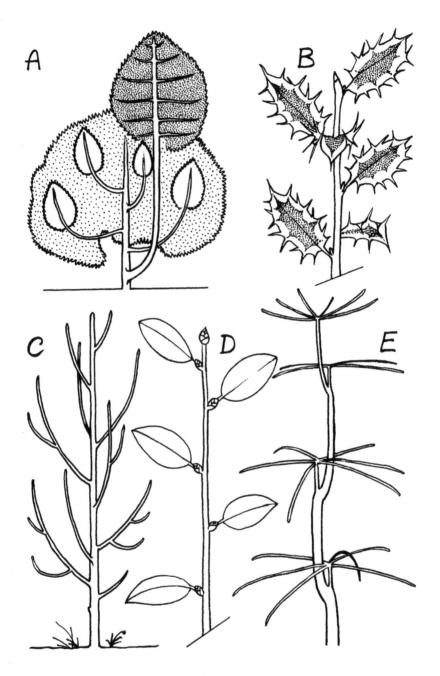

Figure 68. Sorting out mutations. (A) Variegated holly (*Ilex*). Parts of the crown deprived of chlorophyll are shown in white, variegated parts in light gray, and a basal sprout with normal chlorophyll in darker gray. (B) Variegated leaves. (C) A monopodial tree. (D) A leafy branch with lateral buds. (E) A tree with a sympodial trunk.

It is logical that variegated plants, with less chlorophyll than normal plants of equal size, should thus be less vigorous. Moreover, a variegated plant capable of producing normal shoots would consequently tend to overgrow the variegated branches. Anyone experienced with variegated plants knows that it is necessary to carefully remove all normal branches that develop because they rapidly get the upper hand and the plant loses its ornamental character.

Thus branching also appears to be a means by which a plant may remove unfavorable mutations, an idea developed by Klekowski that he creatively applied to plants over their life spans (Klekowski and Kazarinova-Fukshansky 1984a, b, Klekowski 1988, Klekowski et al. 1989). A typical small tree is monopodial (Figure 68C) because the same meristem produces the trunk from the base to its highest reach. The lower branches are at the same time older and younger than the higher ones (please excuse this paradoxical formulation). They are the oldest in real time because they were produced before the higher branches appeared. However, they are biologically younger because their meristems are separated from the egg by fewer cell divisions than is true for the higher branches. A harmful somatic mutation in the meristem that produces the trunk of such a monopodial tree could be present in the higher branches but absent from the lower ones, which are chronologically older but biologically younger. This idea is applicable to a leafy branch with lateral buds (Figure 68D). Lower buds would be less likely to have somatic mutations than higher ones. If the mutations were harmful, these lower buds would be capable of better growth.

The trunks of other trees, described as sympodial, are produced through collaboration of several successive meristems, each one taking over from the previous one. The delayed branches, which produce the trunk in sympodial trees (Figure 68E) such as cocoa (*Theobroma cacao*), pulai (*Alstonia scholaris*), or balsa (*Ochroma pyramidale*), have the effect, if not function, of eliminating harmful mutations accumulated in the apical meristem that produces each segment (Klekowski and Kazarinova-Fukshansky 1984a, b).

It is interesting to extend this idea to the mechanism of reiteration. The crown of a large tree is partly a product of reiteration (Figure 69). The reiterations are both chronologically and biologically younger than the ultimate branchlets of the branches upon which they sit. That would support the role of reiterations in eliminating

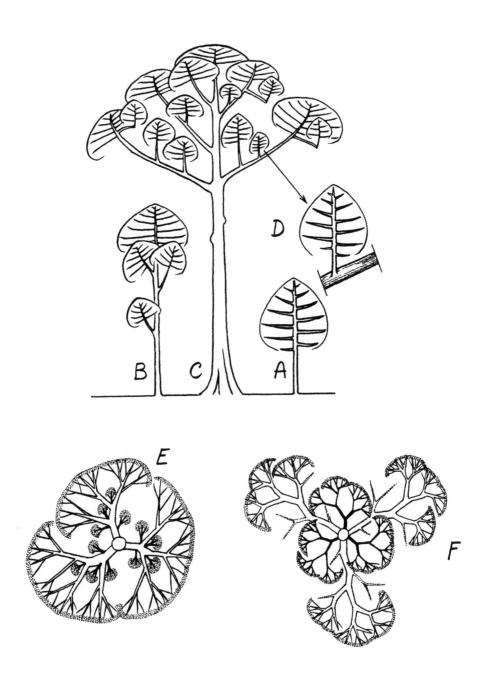

Figure 69. Genetics of reiteration. (A–C) A large tree relies on reiteration to grow, here, *Shorea stenoptera* (Dipterocarpaceae) of tropical Asia (Edelin 1990). (D) A reiteration, younger than the branch carrying it, is comparable to the young plant (A). Reiteration also concerns root systems: Reiterations, shown in black, appear at the centers of the root systems of *Laetia procera* (E, Flacourtiaceae) and *Cecropia obtusa* (F, Urticaceae; Atger and Edelin 1994).

harmful mutations, and thus in the overall revitalization of the tree. Of course, this is a hypothesis for which there is no direct genetic evidence.

Detecting morphological or biochemical criteria for juvenility is one of the directions of agronomic research on woody plants (Perrin 1984). Criteria exist, but it is premature to summarize this subject. Even so, one fact that emerges concerns competition between organs. An animal is an integrated organism, an assemblage of interdependent systems among which competition is excluded by the penalty of failure. This is as Jean de La Fontaine (1621–1695) showed in his fable, *Les Membres et l'Estomac*, the limbs and the stomach. The limbs stop moving as a complaint to the stomach, which is passive and seems to do nothing. Thus the limbs lose their sustenance and their ability to move.

The competitive situation is completely different in plants, organisms that have a repetitive character, little integration, and whose diverse parts may compete with each other. Competition is at the heart of plant function; it is not a rare event but a normal fact of their lives. Competition is lessened when conditions—temperature, light, moisture, soil fertility, space, the biotic environment—are optimal. It is increased by challenges such as drought, cold, etc. An interesting aspect of such intraplant competition is that predatory activity may increase the severity of competition.

Predatory Action

Experts on evolution place great importance on competition between organisms—one of the engines of natural selection—but often forget about competition between parts of the same organism. A judicious observation by Klekowski (1988) incites me to discuss intraplant competition in relation to predation. Genetic plasticity is translated as coexistence of parts or sectors within the plant—leaves, leafy branches, reiterations, older branches—that have genomes more or less significantly different. Following Gill (1986), who found proof through the action of herbivores, many plants should be considered as colonies of sectors with distinct genomes.

Gill and Halverston (1984) reported on an aphid *(Hormaphis hamamelidis)* that attacks witch hazel *(Hamamelis virginiana)*. They observed that the level of attack varied between branches of the same tree, numbered 1 to 58. In winter, witch hazel loses its leaves, the aphids leave it, and they inhabit their alternate host, the black birch (*Betula nigra*). With the return of good weather, the aphids return to the witch hazel trees and feed at the same level of herbivory on each of the different branches as in the preceding year. They repeat this year after year.

These results are consistent with the idea of genetic heterogeneity within individuals of witch hazel, but more than that they show the adaptive value of such heterogeneity. The herbivores attack the parts of the tree that carry "susceptible" mutations, which thus favors the "resistant" parts. Eventually, such plants should become entirely resistant by removal of the susceptible mutations. This is comparable to a variegated plant's removal of the mutation causing lack of chlorophyll and returning to its normal green state. In witch hazel, the herbivores play the active role in intraplant competition; they become the agents that evict the mutants.

Could such a strategy, in the long run, guarantee the protection of a plant against herbivory? Insect predators reproduce sexually in an annual rhythm but the plant remains in place for decades, even centuries. Would not the insect evolve more rapidly than the plant, overcome its defenses, and finally destroy it? How could a tree such as a redwood, which can live 2000 years or more, resist herbivores that have several reproductive cycles each year? For Whitham and Slobodchikoff (1981), the answer lies again in somatic mutation. The frequency of somatic mutation in meristems should always be sufficient to maintain genetic and biochemical diversity in the plant to dissuade predators. Insects, whose diversity and evolution is driven by sexual reproduction, are opposed by the genetic plasticity of plants, resulting primarily from the mechanisms of vegetative growth. These two forms of evolution, transgenerational and intra-organismal, should be equally rapid (Gill 1986), guaranteeing the survival of the two partners.

At this point in the discussion it is important to consider the sexuality of plants and look at the newly fertilized egg. This point of view may seem paradoxical in regard to what we know of the sexuality most familiar to us: ours and that of other animals. I would like to defend the idea that sexuality has its role, along with competition between organs, reiteration, and herbivore pressure, as a mechanism limiting plasticity of plant genomes.

What Does Plant Sexuality Mean?

We have known for a long time that plant and animal sexuality are not the same. Carl Linnaeus (1749), reporting in *Oeconomia Naturae* on his discovery of plant sexuality, wrote, "The genital organs of plants in the plant kingdom are exposed to the eyes of everyone, while the same organs, considered as almost shameful in the animal kingdom, are almost always hidden by nature." In a humorous vein, Marc Oraison went even further: "There is not a festival, celebration, or solemn occasion without flowers. . . . A thank-you, tribute, farewell, apology, arrival, all is transformed by orchids for millionaires or a few violets for poor folk.. . . . Well, flowers are vegetable sexual organs. It would never occur to someone to send the testicles of a bull or the vulva of a cat to show gratitude for some service" (Pelt 1986).

Strange, but such empirical differences should have an objective basis. Even in abandoning this amusing side of this subject, it can be said that sexuality in plants no doubt has the same biological significance as it does in animals, yet is different. In animals, rigorous conservation of the genome and isolation of the germ (Figure 70) ensure not only that gametes are viable but that they will be capable, after fertilization, of forming a viable organism. If a single deleterious mutation occurs in the germ line, the gametes carrying it can be eliminated before fertilization, or the embryo might even die. Such a mutation in the germ line could be beneficial. If so, the gametes participate in fertilization, the genetic novelty is transmitted to descendants, and it is selected by the environment and used evolutionarily. We have before us one of the classical mechanisms of natural selection.

In plants, the situation is profoundly different. Plants do not have a germ line isolated from a somatic cell line (Figure 70). All cells of an aerial meristem may contribute, through their descendents, to the composition of one of the numerous transitory germ lines that plants unceasingly develop in the course of sexual reproduction, at the centers of sporangia, anthers, or ovules. All nonlethal somatic mutations can thus be integrated into the germ. That all viable mutations may be inherited means that germ lines within a plant can contain many more genetic novelties than the single germ line of an animal (Buss 1983). This genomic fluidity, characteristic of

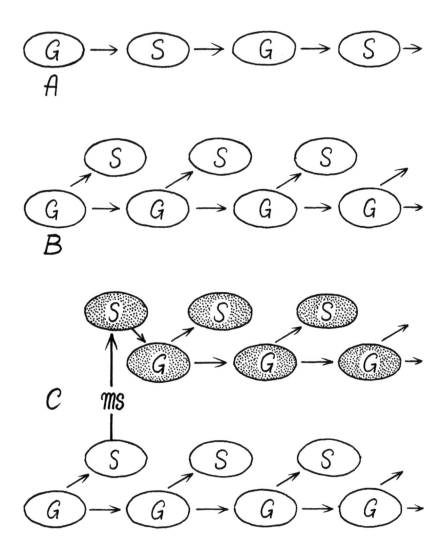

Figure 70. Soma and germ. (A) Schematically, Lamarck and Darwin saw the relationship between germ (G) and soma (S) thusly, whether plant or animal. (B) Weismann (1892) showed that only the germ provides continuity, generation to generation, in animals, the soma being condemned to disappear. (C) Buss (1983) modified Weismann's scheme. Somatic mutations (ms) give rise to variation that, in plants, can be incorporated in the germ.

plants, is an aspect of their adaptability when confronted by environmental vicissitudes from which they cannot escape. However, the high genetic diversity of plant germ lines has a downside: The cells of the varying genomes have not proven their ability to produce a viable organism, and they must be sorted out. That is why plant sexuality has mechanisms for eliminating harmful somatic mutations.

Again, I owe to Virginia Walbot (1985) an extremely interesting idea that the gametophyte, characteristic of plants, could have a role in testing the potential of germ cells for producing a viable organism. The viability of recessive mutant alleles cannot be tested in diploid cells, meaning that lethal mutations could be passed to the next generation. Such a test could not occur before meiosis or after fertilization, but only in between, in haploid spores. The haploid state reveals defects mercilessly; spores containing deleterious alleles are eliminated, and only those producing gametophytes are retained. The gametophytes whose genomes have thus been graded by the mechanism of sexuality eventually produce the gametes, and the cycle is completed. Several points about the test that sexuality provides for plant genomes need further explication, but Walbot's hypothesis remains the only one that attempts to reveal the function of the gametophyte. Its support is growing among biologists (Kondrashov and Crow 1991, Perrot et al. 1991).

Having arrived at this point, and restating that plants favor the genomic diversity that animals reject, it is useful to ask why this is so. In the face of the strict preservation of the animal genome, why is the vegetable genome so fluid?

What Causes Genetic Diversity Within a Plant?

Animals are configured and behave in a way that, in large measure, protects them from the assaults of the environment. Taking advantage of environmental heterogeneity, they actively seek out conditions that favor them. Behavioral plasticity allows an animal to maintain its integrity without having to modify itself.

Plants follow another logic. Because they are fixed in place at a site that they did not choose and from which they cannot escape if conditions become unfavorable, they must be content to elaborate their structure in response to environmental change. The environment

changes ceaselessly, unpredictably as well as regularly, such as alter-
nating between day and night, and summer and winter. Because of
their longevity, plants are at risk of becoming the victims of accidents
—drought, fire, flooding, freezing, lightning—or biological attack—
parasites, predators, pathogens, competitors—because their structure is large, impossible to conceal, alluring, and very vulnerable.

Completely naked and trembling, the great oak outside my window stays there through the winter night, exposed to powerful gusts
from the sea and lashings of hail, while in my bright lodgings, dry
and warm, I have the leisure to reflect on this astonishing contrast.
Doing this, it seems to me that we, the oak and I, express the opposing evolution of plants and animals. Some are free; some submit and
adapt.

By this contrast we search for the significance of genetic diversity
within plants. Because somatic mutations permit rapid accommo-
dation to ecological change (Whitham and Slobodchikoff 1981),
this diversity is in the first instance a strategy for adapting to a fluctuating environment (Walbot and Cullis 1983). Some of the largest
plants can actually live in several environments at the same time. A
tropical forest liana can produce rapidly growing stolons in the
shaded understory while it stretches out under full sunlight in the
canopy. To make this role of within-plant genetic diversity more
concrete, I use an apparently inappropriate example.

The Vertebrate Immune System

Vertebrate animals have at their disposal an immune system whose
purpose is to preserve the integrity of the individual while impeding
the penetration and persistence of foreign elements such as tumors
(ASSIM 1991). The function of this immune system, confronted by
harmful aggressors (antigens) whose number has been estimated to
be 10^9, is comparable to that of a plant that finds itself faced with
harmful ecological aggressors (physical and biological) whose number is beyond calculation.

In practice, the foreign body (antigen) that attacks my individual
integrity can be the blister on which I walk, the spines on my roses,
or the cold virus generously spread by my neighbors on the Métro.
These are the routine aggressors that my body knows well. Invaders
completely unknown to my immune defenses can also appear: a dose

of venom injected by a cone on a coral reef encountered during a vacation to the Maldives, or a malignant cell that invades the lungs after many years of regrettable smoking and multiplies with a vigor more suitable for a better purpose.

Whatever its nature, the antigen transmits a specific signal that arrives at the bone marrow cells, which change into B lymphocytes, carriers of the immune response in the form of a superficial layer of specific immunoglobulins (antibodies) capable of neutralizing the antigens. Thus equipped, the lymphocytes move from the bone marrow into the blood, traversing the capillary network and carrying the immune response that promotes destruction of the antigens, sticking them together with weak but numerous links, a little like Velcro.

The question that arises concerns the specificity of the immune system. How do B lymphocytes synthesize antibody molecules that react so precisely with the antigen, even if the latter is not recognized or is produced by an unknown aggressor? To give an idea of the capabilities of this system, imagine an industry with a chief executive, truly draconian, who forces the factories to manufacture a range of a billion products, many of which do not have models to follow. And production must be prolific and quick—a few days, no more—and delivered almost immediately anywhere in spite of keeping inventory at a minimum. At the same time and without external help, any flaws in the production lines must be tracked down and eliminated under penalty of having them proliferate and destroying the raw material and production capacity. Tolerating only perfection, our executive imposes a norm of zero flaws for all operations (ASSIM 1991).

Susumu Tonegawa (1983) of the Center for Cancer Research at MIT has shown how B lymphocytes satisfy the excessive demands of the chief executive. Pardon me, but the following discussion is somewhat technical yet in relation to what is known gives only a simple picture of immunology. Thus it will appear unpolished to immunologists.

Human chromosomes 2 and 14 (of 23 pairs) carry genes for the synthesis of immunoglobulins (antibodies). These genes, 193 in number, are dispersed on the chromosomes of ordinary cells but are brought together during maturation of the lymphocytes. They are grouped in a random manner and are finally translated in the form of four protein chains that make up the immunoglobulin molecule. The chains themselves combine randomly, forming double combinations. A powerful mechanism for creating diversity is thus in place.

This double combination proceeds only from preexisting genes. Though the combinations may be new, the genes producing them remain the same. The maturation of B lymphocytes is reminiscent of somatic mutation, a mechanism for generating diversity that is even more effective because it extends innovation to the level of the gene. Mutations produced during maturation of the B lymphocytes give evocative names to the genes: V for variable, D for diversity, and J, which assures the joining of the two. These genes are expressed at elevated levels, 10^2 to 10^3 within several days, which immunologists refer to as hypermutation. As Winter and Gearhart (1995) have noted, within two weeks this dizzying phenomenon produces, through mutation and selection, the equivalent of a million years of evolution. The mutations, however, are only conserved and multiplied in B lymphocytes suitable for neutralizing antigens; all others die.

Viewed by a botanist, the immune system of vertebrates provokes comments and questions. For animals, which maintain such strict control of genomic stability, it is remarkable that a category of cells, the B lymphocytes, escapes such control. It is also ironic that these cells are the ones charged with the elimination of genetically deviant cells such as those of malignant tumors. It is no less remarkable that among the diverse strategies adopted by B lymphocytes to neutralize multifaceted and unexpected aggressors, the most efficient is somatic mutation. Of course, this leads us back to plants and an attempt to compare them with the vertebrate immune system.

An interesting resemblance appears between the immune system, which is confronted by a number of unpredictable attacks during the brief life of an animal, and a plant, which also must resist numerous and diverse onslaughts—climate, predators, pathogens, competitors, etc.—during a generally rather long life.

Stationary Lives and Genetic Diversity

Could a plant, in certain ways, be comparable to a vertebrate immune system? I do not push this comparison too far, stating only that both require genetic diversity, leading to resistance against attacks from which flight is not possible. A good way to test this idea is to ask the following question: Do organisms that are fixed in place make an appeal to genetic diversity within the individual? The answer is yes for both plants and animals.

Algae, *Iridaea laminarioides* and *Ascophyllum nodosum*, for example, have an original way of assuring their genetic diversity. Young plants fuse, and the adult resulting from this coalescence quickly forms a mosaic genetic structure (Couvens and Hutchings 1983, Martínez and Santelices 1992). Precocious fusion of sexually produced (and genetically distinct) larvae is also very common. Thus adult sponges "can be ensembles of genetically different cells" (Rasmont 1979). Early fusion of genetically distinct larvae is also common in clonal or colonial marine invertebrates such as corals, ascidians, etc. (Stephenson 1931, Knight-Jones and Moyse 1961, van Duyl et al. 1981, Jackson et al. 1985). Incidentally, if intraplant genetic variability is primarily a strategy for combating the constraints of a variable environment, it could also have another role, contributing to sexual reproduction in some plants, for example.

A self-incompatible tree, one incapable of using its own pollen and thus requiring pollen from another individual of the same species for fertilization, is condemned to sterility if it is isolated and genetically homogeneous. If this isolated tree has a mosaic genetic structure, differences between branches could provide diversity sufficient to allow pollination within the tree (Whitham and Slobodchikoff 1981, Gill 1986). That would help explain mass flowering, covering the entire surface of tree crowns with blooms, which is known to occur in outcrossing tropical trees such as the dipterocarps of Asian rain forests. If the hypothesis of the genetic mosaic is correct, large trees that produce self-incompatible branches could be self-compatible between branches (Gill 1986). Such a hypothesis needs experimental testing, but such tests have yet to made.

Experimentation is much more advanced concerning the resistance plants to herbivory. We have seen that within-tree genetic diversity can contribute to this resistance, with considerable research providing evidence. Resistant sectors within crowns are favored over susceptible regions.

Resistance of Biologists to a Genetics Unique to Plants

Whether phenotypic plasticity (Bradshaw 1965, 1972, Trewavas 1981) or genomic plasticity (Kessler and Reches 1977, Whitham and Slobodchikoff 1981, Buss 1983, Walbot and Cullis 1983, Gottlieb 1984, Antolin and Strobeck 1985, Walbot 1985, Cullis 1986, Slatkin 1987, Klekowski 1988, Charlesworth 1989), such plasticity is advanced more and more often as an important mechanism in plant evolution. There is certainly no consensus on this subject, but it is necessary that plasticity in plants receive due consideration and not be treated as a handicap, as background noise. It is necessary to jump from plant genetics to the psychology of biologists, from resistance to herbivores to resistance to new ideas.

I cast no stones. I was trained according to the animal model, and years later, as I tried to understand the architecture and dynamics of growth of tropical forest trees (Hallé and Oldeman 1970, Hallé et al. 1978), I was often bothered by plant plasticity. It required traveling long distances and observing many trees to determine the precise rules of their morphogenesis. Animals would probably have been more tractable.

Collectively, biologists have poor regard for the idea that plants behave fundamentally differently from animals, that they must, for example, develop within-organismal genetic plasticity under threat of perishing. Such ideas are new to the public and they bother a large portion of the scientific community, trained in the animal model of genomic stability throughout the course of an individual's life: "Plants and animals are only mortal vehicles that transport genes, and their behavior is determined by this special task" (Ruelle 1991). "Individuals are artifices invented by genes in order to reproduce" (Gouyon 1996). "Life has no higher purpose than to perpetuate the survival of DNA" (Dawkins 1995).

I excuse myself from such provocative opinions, borrowed from a sort of divine teleology in which God would be the gene. I reject the idea that people are only a means for the victorious progression of their genes and prefer the idea of Kant that human beings are the ends and not the means to the ends, invested with dignity beyond price. Whatever they may be, these extreme views do not apply to plants; genes would be poorly advised to invest their integrity in organisms who work so hard at producing variation.

Very happily, plant genetic plasticity has begun to attract more realistic points of view. Sachs (1994) wrote that such plasticity is a strategy, not a nuisance; we should consider it a manifestation of the genius of plants by which, in contrast to animals, they successfully manage in a variable environment without having the ability to move (Verhey and Lomax 1993). Do differences in the function of genetic information and in the mechanisms by which genetics supports the organism imply different mechanisms of evolution in plants and animals?

Darwin or Lamarck?

The neo-Darwinian synthesis integrated the separation of soma and germ lines (embryology), classical genetics, and the Darwinian theory of natural selection. Charles Darwin (1809–1882), British naturalist and voyager, elucidated the theory of evolution through natural selection along with Alfred Russel Wallace. The synthesis was established on the idea that evolutionary changes have their origins in natural selection exercised by the environment, benefiting the most fit, that such selection exerts itself on differences between individuals, and that these differences arise from chance mutations. Unfavorable mutations are eliminated along with the individuals in which they are expressed, and favorable ones are conserved. There is practically no relation of cause and effect between environmental constraints and the nature of the changes in the genome; evolutionary progress is, in the final analysis, founded on chance, the environment having only an indirect effect through selection.

A very different mechanism, in which the constraints of the environment directly effect evolution, was postulated by Jean-Baptiste de Monet, chevalier de Lamarck (1744–1829), the French naturalist who proposed the inheritance of acquired characteristics. Environmental constraints would stimulate the appearance of adaptations. Such adaptations, acquired by environmental pressure, would become hereditary and be expressed by the adapted descendants. These acquired characteristics must occur during the life of the individual, following a prolonged effort at adaptation, as an artist acquires expertise or a butcher develops callused hands.

Two fundamental differences separate the natural selection of Darwinism (D) from Lamarckism (L): For D, the role of chance is essential in the generation of mutations and their nature, but not for L. For L, the role of the organism is essential since its adaptation will be transmitted to its descendents. This is only partly true for D, in which the organism is a simple soma, incapable of the slightest influence on inheritance and thus on evolution.

Thus the question of the separation of a germ from a soma is at the heart of the debate on evolution. If such a separation exists, I vote for D; if it does not exist, I prefer L! The quarrel between the partisans of our two heroes shook biology till the end of the 19th century, but it was more often pettily vicious than logically forceful, on both sides, and enmeshed in politics. Capitalism found justification for its economic philosophy in the ideas of D: "Only results matter. . . . The reign of nobility by birth is over. The fruits of the Enlightenment are ripe. Darwin's evolutionary theory is inscribed in descendents." Only the best win! In America, in Western Europe, capitalism and even mercantilism benefited (Reichholf 1993).

On the contrary, in the Soviet Union, Ivan Michurin (1855–1935) and Trofim Lysenko (1898–1976) tried to demonstrate the possibility of the direct hereditary transmission of acquired characteristics, thinking thus to prepare for the advent of communism. A few human generations would require Marxist-Leninist training, then we would see the appearance of a genetically communist humanity. We see the injustice of such absurdity created using the ideas of Lamarck!

When Weismann (1892) established the separation of the soma and germ lines in animals, Lamarckism declined in the face of a new adversary: no longer Darwin, but Weismann. It had already lost ground, and the debate was nearly extinguished by the end of the 1960s. Complete victory for Darwinian natural selection came with the rise of molecular biology.

When I was a student in the 1960s, natural selection reigned triumphantly without exception in evolutionary thought. We were taught that Lamarckian inheritance of acquired characteristics had been completely rejected, that no molecular mechanism existed or could be imagined that would render this type of inheritance possible (G. Ledyard Stebbins according to Landman 1991).

Time has passed, cooling the emotions of the debate and giving us perspective, while the formidable achievements of molecular biology have radically modified the context in which we understand evolutionary mechanisms. In some contemporary scientific literature, it has been shown that inheritance of acquired characteristics may be compatible with the concepts of molecular genetics and that inheritance of acquired characteristics and Mendelian inheritance can coexist comfortably in the realm of molecular biology (Landman 1991).

Are Bacteria Lamarckian?

Bacteria are particularly endowed with inheritance of acquired characteristics; since they are neither plant nor animal, they merit a little detour. They have much to teach us about the two kingdoms that are the focus of this book. Two examples illustrate the ease with which bacteria practice inheritance of acquired characteristics. It is easy to suppress the production of the cell wall that surrounds a bacterium. In *Bacillus subtilis* it is sufficient to mix in an enzyme that depolymerizes the solid constituents of the wall to obtain naked cells, or protoplasts. Their destiny depends on the consistency of the nutritive medium in which they are raised. In a liquid medium, the protoplasts are unable to multiply, and they perish. Placed on a solid medium (agar or gelatin), the protoplasts multiply, surround themselves with new walls, and regenerate the bacteria. The more interesting case is the culture of protoplasts in a medium with a semisolid consistency, a little like cream. In the absence of depolymerizing enzyme, the naked cell characteristic is stable and inherited as long as the medium maintains this consistency (Landman 1991). The acquired characteristic has become hereditary.

Jacques Monod showed that the intestinal bacterium *Escherichia coli* could be induced to make a substance that it does not naturally produce, β-galactosidase. The inducer is MTG (methyl β-D-thio-galactoside), to which the bacteria are exposed at high concentration for a sufficient period. At the end of this incubation, *E. coli* is placed back into its normal culture medium, deprived of the inducer. The bacteria synthesize β-galactosidase, their *lac* operon having undergone a mutation. Their descendants function in the same way; these cells are mutants and the induced characteristic is stable and inheritable indefinitely. This is another example of the inheritance of an acquired characteristic (Monod 1956).

It is important not to play with words. What does *acquired* mean?

To acquire has three meanings: to enter into possession of, to begin to display, or to gain by effort or experience. It is in the latter sense that Lamarck developed his theory of evolution. In the two bacterial examples briefly described, the acquired characteristic is a genetic point mutation, not as the result of any effort or experience. At first glance that would make it seem different from the mechanism conceived by Lamarck. Nevertheless, it is still an acquired characteristic, and a heritable one whose adaptational value is real since its result is viable. We are effectively witnessing a Lamarckian evolutionary mechanism, even if Lamarck—half a century before the discoveries of Mendel—ignored mutations.

Bacteria are thus capable of inheritance of acquired characteristics. Recall the role that bacteria played in the appearance of the first eukaryotic cells (Chapter 3), into which mitochondria (and chloroplasts in plants) were incorporated. It seems that the cyanobacteria that entered and remained in plant cells, where they became chloroplasts, have lost nothing of their ability for inheritance of acquired characteristics. It even seems that they have benefited the plants they inhabit through this ability.

How Weeds Defend Themselves Against Herbicides

An example of one of the ways in which plants demonstrate Lamarckian evolution is provided by the appearance of resistance in weeds to the herbicide atrazine. In the 1960s, this herbicide was used throughout France to remove weeds from maize fields. After several crop cycles, during which the herbicide had the expected effect, resistant weeds began to appear: meadow grass, goosefoot, nightshade, knotgrass, etc. In the United States and Canada, where atrazine was also used, other resistant weeds emerged: senecios, amaranths, mustards, ragweed, etc.

Two different mechanisms were implicated in the establishment of atrazine resistance, depending on the particular weed, as studied by Darmency (1994) and colleagues. The first mechanism is the classical or Darwinian one. Resistance alleles, appearing spontaneously by chance and before any treatment with atrazine, had been maintained in the original populations. The herbicide limited its action to the simple selection of resistant individuals and to the elimination of the others. In untreated populations of velvetleaf *(Abutilon theo-*

phrasti) some plants were sensitive (S) to atrazine; others were resistant (R) or intermediate (I). The allele conferring resistance to atrazine is semidominant, acting by detoxifying the herbicide, and is inherited in classical Mendelian manner. The cross S × R gives a first generation (F_1) that is intermediate. The second generation (F_1 × F_1 = F_2) segregates, 1:2:1 R:I:S. The example of *A. theophrasti* can be extended to most plants, particularly weeds. A unique resistant allele existing prior to any treatment is capable of changing the response of the entire population, favoring the invasion of resistant weeds.

Atrazine resistance in plants may also be the result of far less classical mechanisms. The best example appears to be the white chenopod or lamb's-quarter *(Chenopodium album)*, a weed of maize fields in Europe and particularly studied by Henri Darmency and Jacques Gasquez of INRA (National Institute of Agronomic Research) in Dijon. They distinguished four types of these weeds in the fields that they visited. Fields never treated with atrazine produced rapidly growing weeds described as sensitive (S): 150 g of atrazine per hectare killed them. The homogeneous appearance of the S population of weeds hid a more complex reality. By selfing each of the plants and studying the descendents, a small percentage (less than 10%) was discovered and described as S_p (p for precursor) because 12% of their descendents were of intermediate resistance (I). These 12% required 1000 g of atrazine per hectare to kill them. In some sensitive populations there were about 1% of the plants that were intermediate. Fields that had already been treated with atrazine for several years produced populations of chenopods quite different from those just described. These weeds were not very fast-growing but had enormous resistance to atrazine: 20–40 kg per hectare was necessary to kill them, and they were described as resistant (R).

How does atrazine control the sensitivity of the chenopods? It penetrates leaf cells and moves into the chloroplasts, binding to internal membranes and blocking photosynthesis; the leaves become necrotic and the plants die. Resistance to atrazine is the result of a mutation in chloroplast DNA. The gene concerned with this resistance is *psbA*. A simple substitution of amino acids (from serine to glycine), resulting from a change at codon 264 of the gene, makes the plant resistant to atrazine. The inheritance is not Mendelian; the location of this gene in the chloroplast means that it is inherited maternally (S female × R male = S, whereas R female × S male = R).

The chenopods intermediate in resistance revealed a remarkable response. If treated with atrazine at a dosage a bit less than the lethal 1000 g per hectare, they produced descendents that were 100% resistant, with only massive doses of 20–40 kg per hectare capable of killing them. The mechanism of this transformation (I to R) remains unelucidated, but it occurs in a single generation after exposure to a sublethal dose. This transformation is stable and heritable, disturbing to the agronomist and fascinating to the evolutionist. The critical point is evidently that it is the atrazine itself that determines the establishment of this resistance. A sensitive population is replaced by a resistant population (S to R) in two generations.

The example of the white chenopod is far from unique. We are now aware of more than 55 plant species that have become resistant to atrazine. In a dozen or so, the resistance is the result of the amino acid substitution, serine to glycine. This same mutation confers resistance to atrazine to a range of organisms: cyanobacteria, algae, and vascular plants, including both monocots and dicots. We would *a priori* have expected a diversity of mutations in organisms so genetically distant from each other. This reinforces the idea that the symbiosis between a simple eukaryotic cell and a cyanobacterium, giving rise to the plant cell, was a unique event and not a recurring one.

We also have proof here that the symbiotic bacteria, having become chloroplasts in plant cells, did not cease behaving as bacteria. They easily acquire new adaptational and inheritable characteristics. That they provide plants with this ability is clearly a point of difference between plants and animals.

Must We Choose Between Darwin and Lamarck?

Between Darwin and Lamarck, nothing is certain anymore. Nothing is worked out; that is my personal conviction. My colleague Pierre-Henri Gouyon (1996) has a very different opinion: "To reject [the neo-Darwinian theory] can only be an act of folly, not a scientific decision." Without doubt. Nevertheless, rumors flow and unpublished manuscripts circulate. We have heard that the best journals (those of "A" rank) refuse original articles not because the results are false, but because they support Lamarckian ideas, of which the editors and their review boards will not hear of any compromise. There may be evidence on the side of Lamarck, but it seems that the Inquisition is on the side of Darwin.

In fact, why must we choose between the ideas of Darwin and Lamarck? Why must plants opt for a strictly Darwinian evolution or for another strictly Lamarckian? Since the two mechanisms exist, cannot we admit that evolution could take place by borrowing from one or the other in particular cases?

Plants and animals could also be distinguished by different strategies of evolution. With their strong individuality, stable genomes, germ cells isolated from the soma, and brief lives, animals should evolve by Darwinian mechanisms. The science of evolution was developed by animal researchers; we should understand the importance of Darwinian natural selection in their published discourses.

Who is the troublemaker? Once again, plants. With their symbiotic chlorophyllous bacteria, fluid genomes, open development, potential immortality, and greater opportunism than animals, should not plants also be more Lamarckian? It is good at least to ask this question. Not able to move away when the environment fluctuates, plants should have the opportunity of calling on both evolutionary mechanisms: survival of the fittest and the inheritance of acquired characteristics. Can we push this idea further? Can we imagine that plants are capable of evolving according to a third mechanism, completely original and borrowing nothing from either Darwin or Lamarck?

Geographic Convergence

Frits Went (1971) raised some ideas in the direction of a third mechanism of evolution, attempting to interpret geographic convergences. Went had noticed that in a given geographic area, plants not closely related to each other often share a number of hereditarily determined characteristics. Examples of such convergence have multiplied since Went's time; I begin with three examples at a local scale, borrowed from Barthélémy (1983).

In Mexican deserts, cacti of the genera *Ariocarpus* and *Leuchtenbergia* display an astonishing resemblance to *Agave*. In itself, it is strange that cacti have adopted the appearance of an agave, but it is even more intriguing because these plants live among agaves (Figure 71). Do they derive some sort of advantage from this resemblance?

Even if it desirable to err on the side of caution, it would seem that the answer is no.

In cloud forests covering the flanks of mountains nearly 2000 m high in the Neotropics, tree branches are covered with epiphytes of the bromeliad family (Bromeliaceae). The best known are *Vriesea*, *Aechmea*, *Tillandsia*, and most of all, *Ananas*. These epiphytes are easy to recognize because of their characteristic resemblance to young pineapple plants. As in the preceding example, two epiphytes have assumed the appearance of and grow among the bromeliads though they belong to very different families (Figure 71). The first, *Cochliostema* (Commelinaceae), is in the same family as *Tradescantia*, and the second, *Phymatidium* (Orchidaceae), is in the same family as vanilla and Venus slipper. They manifest a stupefying resemblance to the neighboring bromeliads, "to such an extent that it is difficult to define their exact systematic position in the absence of flowers" (Barthélémy 1983). Here as well, we know of no reason to think that these two imitators derive some sort of advantage from their resemblance to the bromeliads.

In the American Tropics, some species of *Eryngium* (an umbellifer, Apiaceae) imitate bromeliads instead of resembling other species in the genus; one has even been named *E. bromeliaefolium*. These plants live in the alpine tundra of the Andes, where they share the habitat with huge terrestrial bromeliads of the genus *Puya*.

These three examples of convergence, with agaves and with epiphytic and terrestrial bromeliads, occur in small areas, nearly pinpoints, with only a few species implicated, but much larger areas and many more species may also be involved, as shown in the next two examples.

In the course of plant development, whether in a fern, fennel, mahogany, or ash, leaves of the young plant often are simpler than those of the adult (Figure 72). It is interesting to note that the flora of the Mascarene Islands—Mauritius, Réunion, Rodrigues—contains a goodly number of species in which (contrary to the usual situation) seedling leaves are more complex than adult leaves (Figure 72). In order to avoid citing a long list of such plants, I restrain myself to noting that this phenomenon occurs in several common families: Moraceae, Malvaceae, Rutaceae, and Verbenaceae. It seems impossible to explain this juvenile leaf complexity as an adaptation to environmental conditions since the phenomenon can be observed

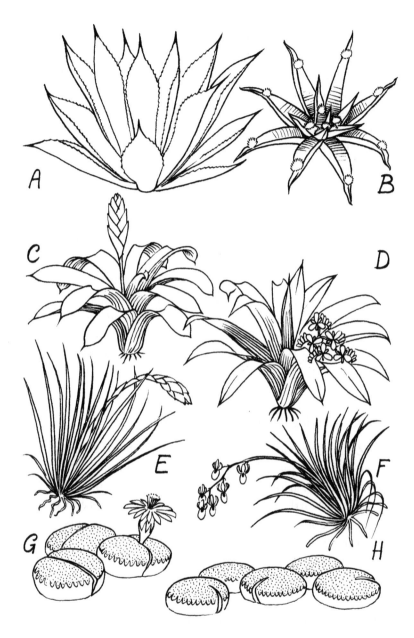

Figure 71. Geographic convergence at a local scale. (A) An agave.
(B) *Ariocarpus agavoides*, a cactus that grows among agaves. (C) An epiphytic
bromeliad. (D) *Cochliostema jacobianum* (Commelinaceae), which grows
among epiphytic bromeliads. (E) Another epiphytic bromeliad, *Tillandsia
stricta*. (F) *Phymatidium tillandsioides*, an orchid that grows with *Tillandsia*.
(G) A living stone, *Lithops* (Aizoaceae). H. Growing among the living stones, a
lily, *Bulbine mesambryanthemoides* (A–F, tropical America; G–H, South Africa).

over a large range of local conditions, dry or wet. Besides, if it were a question of adaptation we would need to see complex juvenile leaves in other regions with comparable climates. But there is none; only the Mascarene Islands have these plant examples.

The case of an arboreal member of the Flacourtiaceae, *Aphloia theaeformis*, is seen more clearly in this context. This tree has a geographic distribution that extends beyond the Mascarene Islands all the way to Madagascar, where it is known as Madagascar tea. *Aphloia theaeformis* only has complex juvenile leaves in the Mascarene Islands; in Madagascar its foliar production is quite normal (Friedmann and Cadet 1976). One geographic region has been touched by this phenomenon, not the other, so there is no obvious adaptational value.

Such is also the case for mistletoes (Loranthaceae and Viscaceae), parasitic epiphytes. Almost 80% of Australian mistletoes closely resemble the tree that they parasitize. On a *Eucalyptus* or an *Acacia* with phyllodes, they produce long, flat leaves. On *Casuarina* or some members of the Proteaceae, they produce linear leaves round in cross section (Barlow and Wiens 1977).

Divaricating Plants of New Zealand

New Zealand provides one of the best examples of geographic convergence on a regional scale. The plants there are the ones that Went studied. Various plants with tiny, oddly arranged leaves grow on both the North and South Islands. Each shrub is a flexible heap of tangled, cohering branches (Figure 73). Numerous branches, thin, long, crooked, and densely branching at right angles, interweave to give these divaricating plants the aspect of gray phantoms. There are 54 species known, belonging to 17 families, and they resemble each other so much that it is nearly impossible to distinguish between them, at least when they are not flowering. Also, it is commonly a single species of a genus that has the divaricating habit (Barthélémy 1983), others having large leaves and a more normal appearance.

Many biologists have viewed divarication as an adaptation to the constraints of the environment: wind, cold, drying of soil. However, this explanation does not hold because these plants grow in a diversity of habitats, including the understory of humid forests, and all in

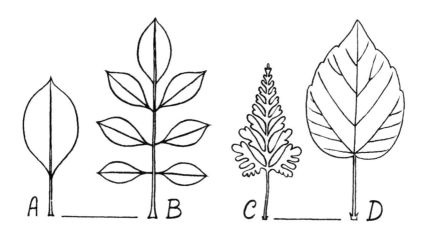

Figure 72. Juvenile foliage forms. In general, the leaf of a seedling (A) is simple when the leaf of an adult (B) is compound. This is true for most legumes. In the Mascarene Islands, the situation is often reversed. The leaf of a seedling *Dombeya populnea* (C, Sterculiaceae) is compound whereas the adult (D) has simple leaves (Friedmann and Cadet 1976).

all, the New Zealand climate is scarcely constraining, especially at low elevations where most of the divaricating plants grow. Dawson (1963) remarked, "In every case the conditions described are not peculiar to New Zealand, but can be found in other parts of the world where they have not been associated with the evolution of the divaricating habit."

Greenwood and Atkinson (1977) hypothesized that divarication could be a defense against moas (Figure 73B), huge herbivorous birds related to ostriches and exterminated by the Maoris in relatively recent times. However, an examination of fossilized moa gizzards revealed that divaricating trees were their favorite food (Burrows et al. 1981). We need to find another idea!

Went's hypothesis had the effect of a thunderbolt in profoundly disturbing the scientific community. Went wrote that such parallel development, or convergence, is the result of transfer of genetic information between quite different species via nonsexual means. In an entirely hypothetical manner, we see how such events could occur. It is likely that we will never know the mechanism of the fortuitous appearance of divarication in New Zealand plants during

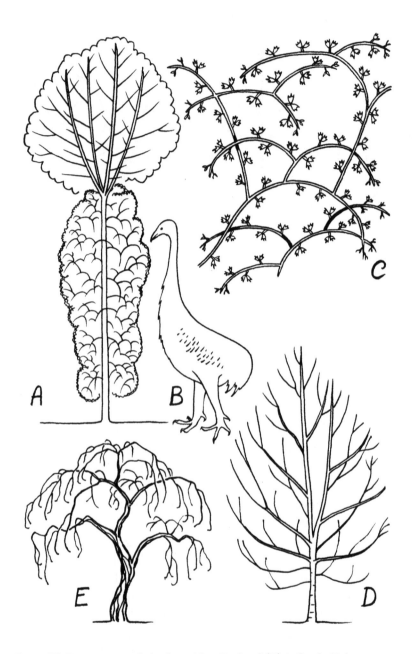

Figure 73. Divaricating plants from New Zealand. (A) A shrub, *Hoheria angustifolia* (Malvaceae). (B) A reconstruction of the moa, *Dinornis*, an herbivorous bird similar to the ostrich. It has been proposed that divarication must have been a defense against moas (Greenwood and Aktinson 1977). (C) A branch of *H. angustifolia*. Divarication is not limited to plants in New Zealand: (D) A normal beech tree, *Fagus*. (E) A divaricating beech in Verzy, France.

some undetermined geologic epoch. It must be the result of some modification of the genome, and thus hereditary. Divarication, not having any adaptational value, but neither being a handicap, persists in the populations of the original species. Imagine that these genes, *a priori* numerous, responsible for the expression of divarication, are grouped together on the same segment of a chromosome.

It is known that certain viruses can become established within the genome of a host, be transferred from one cell to another via mitosis, and from one generation to the next via normal sexual reproduction. Through the mechanism of excision, these viral particles can leave the genome and again become infectious particles. Following an error in excision, such a viral particle could carry with it a DNA fragment from its first host. We can thus conceive of the existence of a modified virus, carrying the information needed for divaricating branches. The introduction of this virus into a second host implies the action of a vector. Mushrooms, nematodes, ticks, piercing and sucking insects such as mosquitoes, even parasitic plants such as dodder, all these could play the role of vector between two plants. The donor and recipient could be two different species.

One of the properties of DNA is its capacity to be integrated into foreign DNA almost immediately if the two come into contact. New genetic information thus transferred into a meristematic cell would be translated in the modified, infected plant and its descendents. This has been called horizontal gene transfer (Pirozynski 1988) and it is how divaricated branching could be spread among New Zealand plants without hindrance from the systematic barriers separating them. This would be a novel evolutionary mechanism (Went 1971). Even if I have given a superficial description, it seems that what is known in the field of virology is not inconsistent with such a hypothesis. Experimental verification is required. Is that too ambitious at present?

The Beeches of Verzy

A vast population of beeches *(Fagus)* is found around the Chapel of Saint Basil in Verzy, near Reims, France. Their strange appearance, recalling that of the divaricating shrubs of New Zealand, probably originated in the admiration, cultivation, propagation, possibly even the collection in a sort of botanical garden, by the Cistercian monks of Saint Basil. It is from the monks who recorded legends about the twisted beeches in the cartularies of the ancient abbey, now destroyed and replaced by the chapel, that we know the beeches were already there in the sixth century.

The trunks, branches, and boughs of these beeches are variously twisted, with many curves or zigzag changes in direction, irregular bulges or fusions, quite anarchic (Figure 73E). The very numerous youngest branches tend to hang down, forming a dense hemispheric vault of thick foliage resting on the ground (Barthélémy 1983). When seeds of these trees are germinated, some of the descendants are tortuous; the characteristic is thus genetic and inheritable. Sown outside the area of their origin, the same fraction of tortuous plants is produced; it is thus a stable characteristic, not influenced by the environment.

Several botanists have proposed that this tortuous characteristic must be propagated virally (Went 1971, Laplace 1977, Barthélémy 1983). That hypothesis is supported by the fact that in the forest of Verzy, neighboring trees are often connected by their roots, which would promote the transmission of a virus. Another surprising fact opens up attractive experimental perspectives. In the 300 hectares of the forest of Verzy, among the 668 twisted beeches there are 13 oaks *(Quercus)* and 3 chestnuts *(Castanea)* that are equally twisted, suggesting the possibility of transmission of genetic information between quite different plants by necessarily nonsexual means. It remains to think up a critical experiment.

Even if it has not received the deserved attention by evolutionary biologists, Went's interpretation of geographic convergence in plants, having received the endorsement of a number of biologists (Pirozynski 1988), has validity. The debate on the dangers of transgenic plants must take into account that plants, unlike animals, do not know how hold onto their genes or, consequently, their transgenes. They seem capable of transferring them to their neighbors even if the latter belong to different species, genera, even families. What would happen if a transgenic rapeseed plant, carrying an herbicide-resistance gene, made a gift of it to the weeds around it? If the gift were made by classical sexual means, then only weeds of the same family (crucifers, Brassicaceae) would benefit. If horizontal gene transfer occurred, then the entire weedy flora of the region could become resistant to the herbicide.

We are likely deciphering an old evolutionary mechanism that may find conditions favorable for developing on a global scale in plants. Viral transfers of genetic information between different animal species have been documented (Martin and Fridovich 1981), but the mechanism seems reserved nearly exclusively to plants.

Mimes and Mimicry

In the case of mimicry, geographic convergence may lead to rapid evolution. We distinguish two types of mimicry: one in which the mime tries to escape notice, and another in which it tries to pass as another organism (Pasteur 1995). Animals have practiced both types, and examples are not uncommon. A gray spider rests on a tree trunk covered with lichens, a seahorse *(Phyllopteryx)* moves passively along with the algae surrounding it (Figure 74C). These are attempts at camouflage.

In contrast, when a female firefly *(Photuris)* produces light signals imitating those of another firefly *(Photinus)* and a male *Photinus* is attracted and then devoured by the *Photuris*, the larger organism, it is an example of homotypy. An anglerfish (family Lophiidae) wriggles a fishing filament in front of its mouth, nearly invisible but terminating in a lure in the form of a worm, shrimp, or young fish. Attracted by this moving bait, a fish swims toward it and is engulfed by an enormous mouth, another victim of homotypy (Figure 75).

Batesian mimicry [Henry Walter Bates (1825–1892), a disciple of Darwin and student of Amazonian natural history, discovered this among butterflies] is often utilized by animals, from insects to vertebrates. We know of many toxic animals that cannot be consumed with impunity, or even handled. The *Heliconius* butterfly (Figure 51, Chapter 4) is toxic to birds, which avoid eating it; the *Dendrobates* frog is toxic to predators; if you grab it, you quickly experience the horribly painful poison contained in the mucus covering its skin.

What pushes animals to become toxic if it does not prevent predators from eating them? What good is it to produce a poison if it threatens only the life of the predator doing the feeding? We understand why such animals are usually brilliantly colored: It makes them recognizable, and avoidable. *Heliconius* and *Dendrobates* are superb animals, noticeable at a glance. This is where Batesian mimicry takes over. In tropical America where this story unfolds, all butterflies

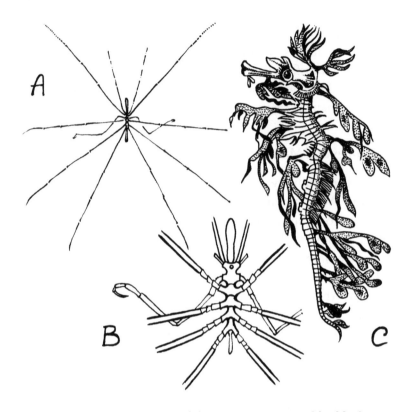

Figure 74. Camouflage in animals. (A) A marine pycnogonid (width about 2 cm), related to spiders and nearly invisible among the vegetation. (B) The body of the pycnogonid is so thin that digestion takes place partially in its legs. (C) *Phyllopteryx*, an Australian seahorse. Its olive green color helps it hide in tufts of algae and escape predators. The relationship between surface area and volume in these two animals is similar to what we observe in plants.

resembling *Heliconius* are, from this fact, protected from predation. The more complete the resemblance, the more efficient the protection. On the other hand, butterflies that bear little resemblance to *Heliconius* have a high probability of being eaten by birds. In time, with the system of variation and selection playing its role, species of *Heliconius* find themselves surrounded by other, nontoxic species mimicking them (Figure 51). The same is true for *Dendrobates* and other frogs.

Following its discovery by Bates in 1862, mimicry was not at first recognized as an important phenomenon in animal biology. Fabre considered it "child's play and did not have words scornful enough

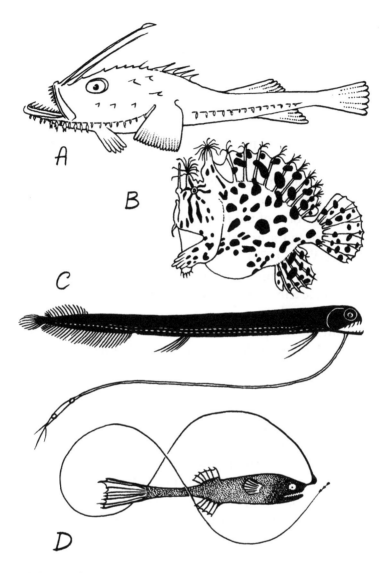

Figure 75. Lures, homotypy, and camouflage in fishes. (A) The anglerfish *Lophius piscatorius* is 2 m long as an adult. Fishing filaments bearing lures and fragments of skin are plant-like in appearance. (B) The frogfish *Antennarius occidentalis*, only 8 cm long, lives in the Sargasso Sea and is mistaken for seaweed because of its lures, immobility, coloration, and skin fragments. (C) The ventral light organs of *Eustomias australanticus*, a dragonfish of great depths, allow it to escape unnoticed by predators coming up from below because it thus merges with the faint glimmer of light coming from the ocean's surface. (D) The whipnose anglerfish *Gigantactis macronema*, another fish of great depths, has a fishing filament three times the length of its body (after a model in the Natural History Museum, London).

to condemn those who pretended to observe it" (Pasteur 1995). Time has passed, however, and mimicry is recognized as one of the engines of animal evolution, especially in insects, and hundreds of publications have been devoted to it. It is legitimate to ask if it also could be an engine of evolution in plants. Are they endowed with mimicry?

Plants know how to practice camouflage, of which the best examples are found among the stone plants of South Africa. *Lithops* (Figure 71G) does not resemble just any pebble; these living stones are of the size, form, color, even having the fine cracks, of surrounding stones. Herbivores of the region, ostriches and ungulates, do not notice *Lithops* and search elsewhere for food that is growing under their feet (Wiens 1978). Plants also know how to produce homotypes, the best example being orchids of the genus *Ophrys*, whose flowers imitate female bees in exact detail (Chapter 4), assuring their pollination by male bees.

In contrast, few plants have evolved mechanisms of Batesian mimicry. Many toxic plants are recognized visually by herbivores and carefully avoided. That is true of the Curaçao milkweed *(Asclepias curassavica)*, which is not grazed by cows and tends to invade pastures in many regions of the American Tropics. We wait for the mimics that should appear nearby, edible but benefiting by their resemblance to the toxic model, but there is nothing. This type of mimicry for which the purpose is perfectly clear, so frequent in animals, does not seem to exist in plants. Examples are only found among plant organs: A passionflower possesses stipules mimicking the eggs of a butterfly; seeing that their place is taken, the butterflies avoid laying eggs there (Gilbert 1991). A clover possesses a mixture of edible and toxic leaves, the latter protecting the former (Till-Bottraud and Gouyon 1992). The authors give these examples a Batesian interpretation, but at the very least we are surprised by their fragmentary character.

Why is there not a plant equivalent to the examples of *Heliconius* or *Dendrobates*, a nontoxic plant escaping predation thanks to its resemblance to a toxic plant? We can only hypothesize. Plants protect themselves by their chemical composition and not by their visual appearance. In fact, insects, their main predators, choose them on the basis of chemical, not visual, criteria. Under these conditions a change in external appearance would mean nothing, and Batesian mimicry in plants should perhaps be studied at a chemical level (Wiens 1978).

If Batesian mimicry makes sense for animals—highly integrated beings whose vital organs must be protected—it would not make sense for plants with their poorly integrated structure and whose organs have a reversible nature. Plants survive thanks to their indeterminate embryogenesis (White 1984).

We close this digression on mimicry. It has led us to the conviction that it is mainly an animal practice in which plants participate only a little. The balance sheet for the evolutionary mechanisms at work in the two kingdoms reveals some profound differences.

Two Different Classifications

The systematics of plants and of animals differ in their methods, objectives, and results. We observe and understand more easily what is familiar to us rather than what is strange. A European who arrives in Africa or an Asian who moves to Europe often has difficulties in recognizing new acquaintances. Everyone looks the same, they say. The same phenomenon colors our understanding of plants and animals because we belong to one group and are strangers to the other. Young children can distinguish the large groups of animals—mollusks, insects, fishes, birds. However, they lack the advantage of experience for recognizing groups of similar importance among plants—dicotyledons, bryophytes, pteridophytes, gymnosperms.

If we were plants, perhaps we would see a teeming mass in animals, noisy and relatively homogeneous. Attempts to classify them would be the business only of specialists who would pass their time establishing distinctions between beasts on the basis of characters that we would find subtle, even pointless: to swim, fly, or creep; two, four, or six feet. Who could attribute any importance to such tiny details?

Can we avoid bias in the study of plants and animals? In the 18th century there was the "compleat naturalist," Carl Linnaeus. Was he objective when he established the binomial method used today for both plants and animals? Certainly not. Linnaeus always preferred plants, which reminded him of his parents' garden in Råshult, Sweden. His biographers, Heller (1964) and Blunt (1971), noted with regret that his work on animals had been, shall we say, too hasty.

Both animal and plant systematists throughout the world use the Linnaean practice of binomial nomenclature in Latin. However, they have not succeeded in agreeing on the manner of indicating the authors of their binomials. I give two animal examples: (1) *Papilio machaon* Linnaeus, 1758; this insect received the generic name *Papilio* and the specific name *machaon* from the great naturalist on the indicated date, and it is still used. The machaon, *P. machaon*, belongs to the family Papilionidae in the order Lepidoptera. (2) *Zonocerus variegatus* (Linnaeus, 1758); the stinking cricket had been described by Linnaeus in 1758 under the name *Gryllus locusta variegatus*. In 1873, the Swedish zoologist Carl Stål named the genus *Zonocerus* and included in it as a species what Linnaeus had described as a variety of another species. Neither Stål's name nor the date is given. We know only that Linnaeus is the person who first described the cricket, under another name, the parentheses indicating that. This cricket belongs to the family Pyrgomorphidae in the order Orthoptera. For comparison, I give two examples from plants: (1) *Ruta graveolens* L.; rue received its binomial on an unspecified date and by an author known only by the initial L. (for Linnaeus). The name has not been changed since. Rue belongs to the family Rutaceae in the order Sapindales. (2) *Pycnanthus angolensis* (Welw.) Warb.; the ilomba tree was first named *Myristica angolensis* by Friedrich Welwitsch in 1862. Thirty-three years later, Otto Warburg judged that it was misplaced in *Myristica* and decided rather to put it in the genus *Pycnanthus*. It carries, for our reference and generally in an abbreviated form, the names of the botanists who have successively studied the plant, but in contrast to animals the names are not dated. Ilomba belongs to the family Myristicaceae in the order Magnoliales. I suggest that in 2007, the tercentenary of the birth of Linnaeus, that these two systems be unified by indicating both the names of the describers and the dates of their work. We can also use the same suffixes in the two kingdoms for orders and families, making it simpler for nonspecialists.

Systematics in practice testifies eloquently about our partiality. Animal systematics is given an elevated goal: the understanding of evolution. Unlimited by practical considerations that would shackle its freedom, enriched by examples from embryology, anatomy, paleontology, and powerfully assisted by cladistics and molecular phylogeny, it succeeds in establishing the links of "vertical" parentage to reconstruct evolution. The latter is adequately described by the metaphor of the tree (P. F. Stevens 1984).

Behind such success, there is, it must be said, the collective attention that we give to that which affects us deeply. We feel that medicine will benefit from progress in our knowledge of animals. Consequently, financial means and human energy are placed at the disposal of the zoologists. Despite the exceptional opportunities for experimentation they provide, plants have not often excited the collective interest of researchers; this is the nature of things.

Historically, the classification of our green cousins has never had much ambition behind it; it developed above all to provide a service to horticulture, agronomy, phytotherapy, and teaching. Springing from too exclusively an applied objective, the origin of plant systematics has proved inauspicious. To facilitate the easy and reliable identification of economic plants—for food, ornamentation, and above all, medicine—plant systematics has always tended toward the production of keys and floras. In the writing of these latter works, external morphological characters have always been favored, judged as sufficient for establishing useful classifications.

Beyond the fact that it has not been considered as an end in itself, plant systematics suffers from several handicaps. Embryology has had nothing decisive to bring to it apart from the number of embryonic leaves used to separate the monocotyledons—narcissus, marram grass, arum—from dicotyledons—balsam, tulip tree, daisy. Was it justifiable to use a difficultly observed embryonic feature to separate two biological entities differentiated by other, obvious characters?

Other handicaps are still more serious. It is rare that fossils shed light on the phylogeny and classification of plants. The notion of a species in plants is borrowed from animals even though the latter live according to a quite different logic. Scientific collections of plants (herbarium specimens) almost always consist of fragments, as if we preserved only the wing of a fly or a hair from an elephant's tail. Beyond that, one essential part of plants, the root, is underground and thus poorly known.

Thus handicapped, what can be done about plant systematics? It is content to regroup plants with a certain number of characteristics in common together as the same taxon, which is to say, those having attained the same evolutionary status. Given that the choice of characters judged important varies from one group of plants to another, the taxa thus established are "horizontal" and, for those of the highest levels, notoriously composed of several evolutionary lines joined together—we say that they are polyphyletic (P. F. Stevens 1984).

Plant evolution is generally only discussed in terms of relationships between actual taxa, evidently not satisfactorily. Hope founded on cladistics and molecular phylogeny has been partially deceiving; these methods work satisfactorily only if they are used by a specialist who knows the group examined particularly well. The expert thus chooses the criteria of analysis in such a way that results conform to the classification previously predicted. These methodological problems have been discussed (Funk 1981), and there are no really good phylogenies for plants, useful over long periods of geologic time, such as we have for cephalopods, horses, or humans.

Although fortified with results from more recent techniques, it would seem that plant systematics has only a few means at its disposal. At least that is what leaders in scientific research think. Who should be surprised? They are all primates! Replace them with great trees and things would be different. I agree with my colleague Peter F. Stevens (now at the University of Missouri, St. Louis), who has written extensively on the disappointments of systematic botany as practiced from 1690 to 1960. He wrote lucidly of "the relative ease with which man could distinguish between organisms more like himself" (P. F. Stevens 1984).

[I have to disagree with the author on his comments about plant systematics but can understand his position given the poor state of the discipline in France, where plant systematics has lagged behind that in the United States, Canada, and much of the rest of Europe. French systematists are limited to the largest natural history museums, whereas in the other countries, plant systematics research continues in colleges and universities in addition to museums and governmental institutes. A principal reason for advances in research in plant systematics is the introduction of techniques for analyzing DNA sequences in cell organelles and the nucleus. These molecular techniques, coupled with the concepts of cladistics put to use in computers, have resulted in new insights about the relationships and evolution of plants. A dramatic example of this work is the three-gene molecular phylogeny of the angiosperms produced by Soltis et al. (1999). —translator's comment]

I conclude that the two evolutions, plant and animal, are intrinsically different. The genealogical metaphor of the tree, describing the evolution of animals, does not work for plants. In plants, hybridization and convergence translate into fusions and anastamoses between the branches such that the tree is not a tree, but a network (Figure 87, Chapter 6). Plant evolution is fundamentally reticulate, which also explains difficulties in plant cladistics (Joël Mathez, personal communication).

CHAPTER 6

Of Other Living Beings

Clairvoyant like artemisia, [the vervain] makes the sick cry who will heal and makes the dying laugh. It is enough to enter the room while holding the herb. It is not that vervain is a bad joke. It is gathered while walking backward; it is of another world.

PIERRE LIEUTAGHI, *La Plante Compagne*, 1991

Between mushrooms and volcanoes is a larval state of water wanting to become fire.

JACQUES LACARRIÈRE, *Lapidaire: Suivi de, Lichens*, 1985

I like trees because they seem more resigned to the way they have to live than other things do.

WILLA CATHER, *O Pioneers!* 1913

IN THE PRECEDING PAGES, *animal* and *plant* have designated freely moving animals, and plants fixed in place. The words animal and plant are used comparatively for the living beings that appear in the foreground of our landscapes, those who are directly and easily observable by everyone. Reality is much more complex, however, and life is not limited to aphids in a garden, ponies in a meadow, or a jaguar in a forest. I wish to embellish the comparison by mentioning some other biological examples. They are certainly less accessible and less familiar, but they improve our understanding of the duality of plants–animals while sidelighting some most revealing things.

Fungi

Fungi must be mentioned. The personality of these beings, and that of the kingdom, lies in a curious mixture of animal and plant characteristics. Considered plants for a long time, fungi are similar to plants in their incomplete multicellularity (Figure 76), in the absence of a boundary between soma and germ, in their cytoplasmic current (or cyclosis; Chapter 3), in their production of defensive toxins, and finally in their immobility—with some exceptions. In contrast, they do have several animal characteristics, particularly a lack of chlorophyll and a heterotrophic mode of metabolism, depending on preexisting organic materials. Their digestion is not internal as it is in animals. Fungi excrete powerful enzymes that reduce organic food to small molecules. The latter are then transported into the cytoplasm across cell membranes. Chitin, which they synthesize for their cell walls, is another animal characteristic, as is the frequent presence of centrioles and asters during cell division. Mobility is a characteristic to consider if one includes slime molds, the myxomycetes, among the fungi, which Margulis and Schwartz (1988) refuse to do. The mobility of the cellular slime mold *Dictyostelium discoideum* when it adopts the form of a slug was described in Chapter 2.

WHEN IT COMES to mixing animal and plant characteristics, fungi are trumped by corals. Incontestably animals, corals have such numerous and highly visible plant characteristics—a lack of mobility, branching, reaching toward light, etc.—that they were confused with plants till the end of the 18th century. Only in 1723 did the naturalist Jean-André Peyssonel suggest that corals are animals in a communication to the Academy of Sciences, Paris. This assertion was rejected with derision, and Peyssonel abandoned all scientific research (Goreau et al. 1979). Later, he was proved right, and the animal nature of corals was fully recognized even if many of them bear plant names: *Pectinia lactuca, Acropora hyacinthus, Pavona cactus* (Harper et al. 1986).

What characteristics are peculiar to sedentary life? What forms and functions does the absence of mobility make obligatory? Corals, better than plants alone, provide part of the answer. Why better than plants? From our previous perspective, creatures living fixed in place are almost all plants, and we risk confusing characteristics of this mode of life with those of vegetable life. In order to distinguish between the two, comparing plants and corals seems a good method. In any case it is very agreeable work, requiring frequent comings

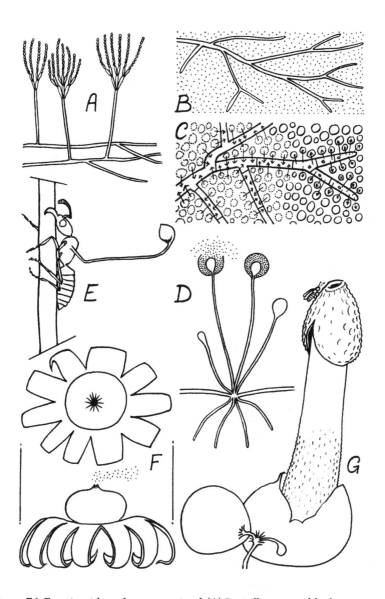

Figure 76. Fungi, neither plant nor animal. (A) *Penicillium,* a mold whose vertical branches, producing spores, are 200 μm high. (B) The filaments or hyphae of *Penicillium* lengthen and branch in a nutritive medium, for example, plant tissue. (C) Secretion of extracellular enzymes (right) is followed by absorption of the products of digestion (left) by the hyphae. (D) *Rhizopus,* another mold, with sporangia 2 mm high. (E) *Cordiceps,* a fungus parasitic on tropical ants. Following the death of its host, it produces a reproductive structure 2 cm long. (F) *Geaster,* 3 cm in diameter. (G) *Phallus,* 20 cm high; the upper part, where spores are produced, diffuses an odor that attracts flies. *Geaster* and *Phallus* are forest fungi.

and goings between the forest and the reef, separated only by the dazzling beach and the jade lagoon. Even if study is its primary concern, scientific research can be a source of much pleasure.

Trees and Corals

The constructors of reefs, hermatypic corals, comprise species of a diversity, longevity, size, and geographic distribution that make the group ideal for comparison with trees. This analysis should be done using as many and as varied characters as possible. Plants other than trees, and other animals fixed in place and aquatic, provoke an even more interesting debate.

The first similarity between corals and trees appears in the capture of solar energy by reef-building corals that live symbiotically with zooxanthellae. These unicellular algae live in certain cells of the coral (Figure 77C) and their density—a million zooxanthellae per square centimeter—gives the corals a golden brown color. The products of algal photosynthesis benefit the coral, supplying its essential energetic needs (Schumacher 1977). For corals, as for trees, solar energy is a vital necessity. This explains their growth in height and competition for light.

Photosynthesis by symbiotic algae is not the only source of food for the coral. The polyps that compose it are carnivores and capture small planktonic animals. The similarity to photosynthesis persists: In one case as well as the other, living fixed in place means satisfying the appetite with food brought to the captor; the captor does not move toward the food. In both cases, there is weak energy flux, making the formation of large surfaces to capture prey indispensable. The coral, like the plant, is a huge fixed surface, and for the same reasons as in plants (Chapter 2) these surfaces are for the most part carried on a system of axes.

Growth reveals other similarities (Figure 77). Corals grow in length through the function of growth zones that are equivalent to plant meristems, with polyps situated at the tips of the axes. Growth is indeterminate in corals and plants, and prolonged throughout life. Increases in diameter are also comparable. Underneath the base of the polyp, between it and the skeleton of the coral, a zone of calcium crystallization occurs. The crystals of aragonite formed here are deposited at the surface of the skeleton, whose thickness in-

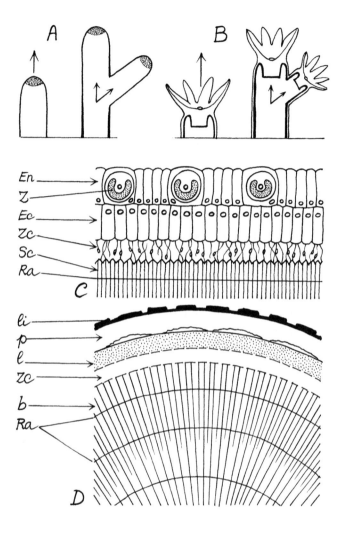

Figure 77. Growth of corals and trees. (A) Lengthwise growth and branching of a tree. (B) Lengthwise growth and branching of a coral. The polyp, equivalent to the plant meristem, is elevated by the products of its own activity. (C) Growth in thickness of a coral. A layer in the epidermis called the gastrodermis (En) contains zooxanthellae (Z). A layer of epidermis above the gastrodermis is not shown. Underneath the gastrodermis, in a separate layer of epidermis (Ec), a zone of crystallization (Zc) augments the thickness of the calcareous skeleton (Sc). Increase in thickness is marked by growth rings (Ra), which show an annual rhythm. (D) Growth in thickness of a tree. The cambium (Zc) is located between the inner bark (l) and wood (b). Increase in thickness is marked by growth rings (Ra), which show an annual rhythm. Toward the exterior of the trunk, parenchyma (p) and bark (li) play a protective role.

creases. In a similar fashion, a tree's cambium produces new cells impregnated with lignin, adding to the wood already formed, whose thickness increases. In either case, the volume of aragonite or wood increases with time and is excremental in nature (François Feer, personal communication). In corals as in trees, annual growth layers, or rings, transcribe the rhythmic character of the thickening, the outermost ring being the youngest.

Coral Architecture

Like trees, corals are usually capable of branching, which is to say that the number of their axes increases with time, optimally filling a volume for energy capture (Figure 78). As always, there are exceptions, such as species in both groups that do not branch, the coral *Fungia* being the equivalent of an oil palm or a cauliflower. In both corals and trees, we find radial symmetry, the result of vertical growth at right angles to the substrate, and polarity from base to summit. This comes from the fact that the basal portions are oldest and the highest parts youngest.

At the scale of the entire organism, neither trees nor corals display anteroposterior or dorsiventral polarity, which are characteristics of active mobility. In contrast, dorsiventrality occurs at the scale of the organ—leaf, lateral polyp of *Acropora*—or group of organs—branches of kapok or fir, horizontal branches of *Acropora*—and this partial dorsiventrality comes from the need to capture energy in both cases (Figure 78).

Much research has shown the existence of architectural models in trees (Hallé and Oldeman 1970, Hallé et al. 1978, Edelin 1990). These schemes of branching are generalized, observable in species not necessarily related taxonomically. Dauget (1991) has shown that architectural models also occur in corals, and in some cases they are identical to those in trees (Figure 79).

Reiteration and colonialism are common to both trees and corals. Reiteration (Oldeman 1972), complete or partial repetition of the architectural model, was discovered in trees, then found in corals (Figure 80; Dauget 1985). The colonial nature of corals has been known a long time, in which the individual unit is the polyp. On the contrary, the colonial nature of trees is a much more recent concept (Oldeman 1972). In the colony of a tree, the elementary unit is the

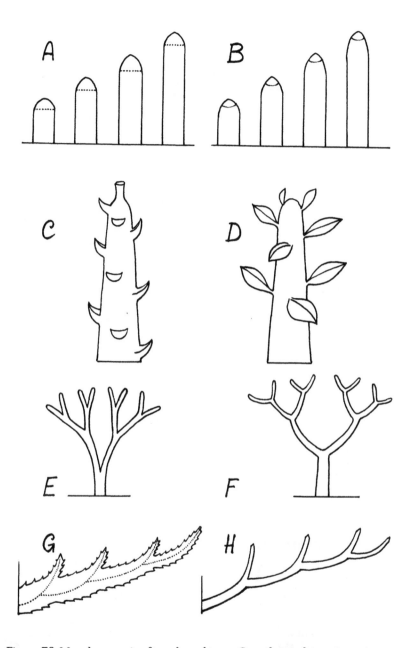

Figure 78. Morphogenesis of corals and trees. Growth is indeterminate in both. Growth zones in corals, formed by numerous polyps (A), act like meristems in plants (B). Lateral polyps of *Acropora formosa* (C) are dorsiventral like the leaves of a tree (D). Branching of the coral *Lobophyllia corymbosa* (E) is comparable to that of a lilac (F). Horizontal branches of *A. hyacinthus* (G) have the same architecture and growth dynamics as the lower branches of many tree species (Dauget 1991).

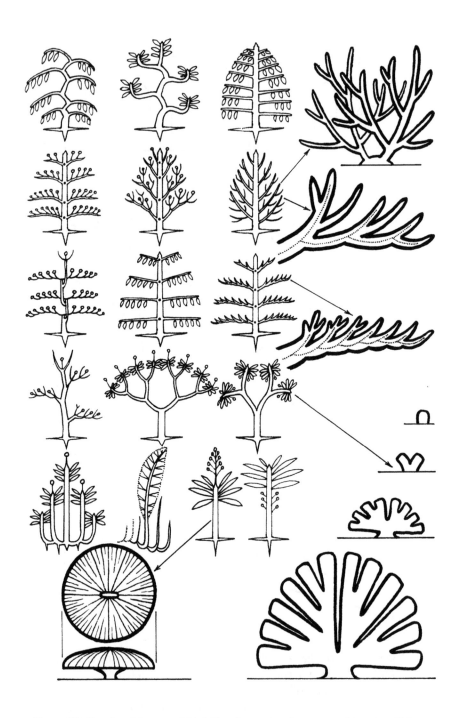

Figure 79. Coral architecture (thick lines) compared to tree architecture (thin lines). Branched corals form architectural models identical to those of trees in the examples shown (Dauget 1991).

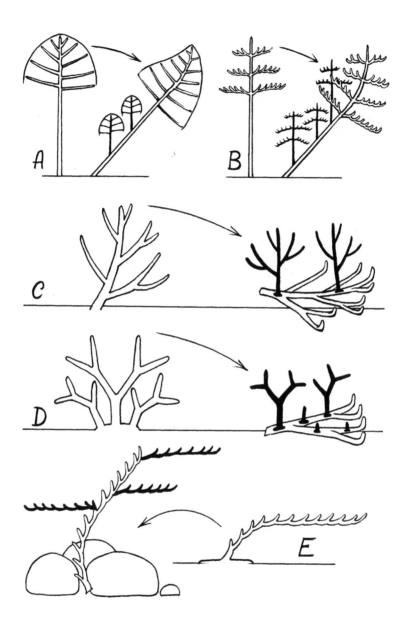

Figure 80. Reiteration in trees and corals. (A, B) Two trees have been placed in an inclined position, by a windstorm, for example. Reiterations appear, restoring verticality (Oldeman 1972), while vertical growth reestablishes the top of the tilted tree. (C) A living branch of the coral *Acropora nobilis* is placed horizontally. The coral reiterates like the tree, and vertical growth is reestablished at the tip of the battered branch. (D) The same phenomena are seen in the coral *Porites cylindrica*. (E) The coral *A. hyacinthus* normally grows horizontally. Placed vertically, it reiterates horizontal branches (Dauget 1985).

reiteration (Edelin 1990). Reiteration and colonialism are opposites of the unitary nature of free-living, mobile animals, which are unable to repeat the sequence of morphogenesis.

Longevity of organisms is directly linked to their unitary versus colonial nature: No freely moving animal seems capable of living more than three centuries, a record attained by some tortoises. The longevity of plants is dramatically superior, and some among them are potentially immortal (Chapter 2). The same is true for reef-building corals, which have immense, even geologic, life spans. Their growth accompanies that of volcanic islands as they successively form fringing reefs, barrier reefs, and finally atolls. They are potentially immortal (Jackson et al. 1985). The same duration of life (13,000–15,000 years) has been proposed for corals and the most ancient trees.

If part of the organism is amputated, a leg, for example, most freely moving animals must be content with simple healing, with a few exceptions: planarians, starfishes, crabs, lizards, electric eels. On the contrary, the axes of corals and the leafy branches of plants normally regenerate. This is a rule with very few exceptions (Figure 81).

The English ecologist John Harper (1977), interested in analogies between plants and corals, proposed a terminology common to the two. He called the product of sexual reproduction a genet, which thus represents a single genotype. During development and growth, the genet is dispersed into a large number of ramets. Asexual reproduction, which produces the ramets (nonexistent in freely moving animals except in special experimental situations), is widely distributed among animals living fixed in place—hydras, corals—as well as plants. It can be spontaneous, such as the budding of corals and the suckering of plants (Figure 82), or artificial—stem cuttings, commonly used in plants, is practiced with equal success in corals (Gérard Faure, personal communication).

Returning to the Idea of the Individual

A free-living, mobile animal is an individual, signifying that it is not divisible without causing its death. Fabre (1996) asked, "Who recommends swinging the hatchet into the cat to divide it into two equal parts, in the hope that the two moieties will continue to live and form two distinct animals henceforth each with its own existence?" Reit-

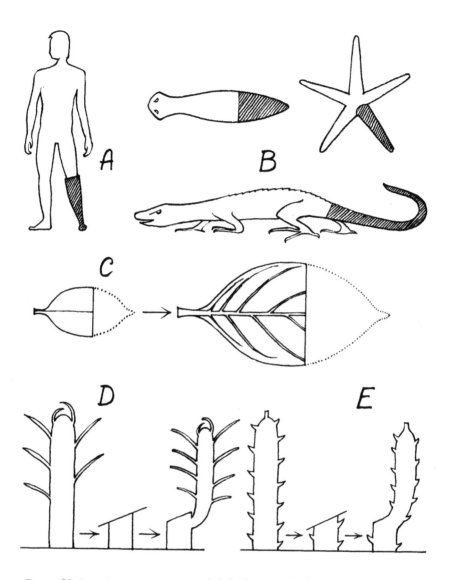

Figure 81. Reactions to amputation. (A) If a human's limb is amputated, he or she must be content with simple healing of the wound. This it the generally true for free-living animals. (B) Exceptions exist but not many: In a starfish, planarian, or lizard, amputation is followed by regeneration. (C) In a young leaf, amputation is not followed by regeneration and the adult leaf is incomplete. This type of reaction is characteristic of most dorsiventral or bilateral organs in plants. In contrast, amputation of a radially symmetrical axis in plants (D) as well as a branch of coral (for example, *Acropora*, E) is followed by regeneration, and the accident is quickly made undetectable.

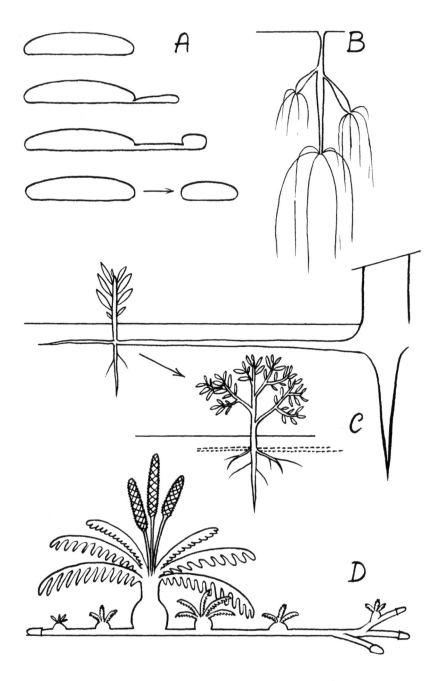

Figure 82. Asexual reproduction in nonmoving organisms. (A) Budding of the coral *Fungia fungites,* Seribu Islands, Indonesia. (B) Budding of a freshwater hydra. (C) Root suckering of a tree. (D) Root suckering of *Hydrostachys myriophylla* (Hydrostachyaceae), Zongo Falls, Democratic Republic of the Congo (Zaire).

eration, colonialism, regeneration after trauma, spontaneous or artificial asexual reproduction—all are mechanisms incompatible with the idea of the individual. Even if they are not colonial, such as the annual herb *Arabidopsis* or the coral *Fungia*, these organisms, fixed in place, are not killed if divided into two equal parts. That is true, *a fortiori*, for trees and colonial corals. Physiological and genetic considerations reinforce the necessity of abandoning the concept of the individual for beings who live fixed in place.

An immunologic defense system, more or less perfected in them, protects the individual identities of free-living animals. It enables them to overcome attack by parasites, heal wounds, and eliminate diseased cells, whether deadly or genetically deviant. Neither trees nor corals have true immune defenses. There are mechanisms of recognition of self and nonself, still little understood, that appear to be more effective between different species than between different individuals of a single species. Two okoumé trees *(Aucoumea klaineana)* or two *Macaranga* trees can fuse by their roots, but we never find fusions between an okoumé and *Macaranga*.

Using themselves as a substrate, trees and corals are capable of recolonizing their own dead tissues, covering them up with living tissue (Yves Caraglio, personal communication). We do not know if crown shyness (Ng 1977) functions as a mechanism of recognition of self and nonself, but it has been established that this shyness, first observed in trees, also occurs in corals (Figure 83).

In plants, powerful and diverse chemical defenses in the form of secondary metabolites—polyphenols, anthocyanins, alkaloids, flavonoids, etc.—often assure protection against herbivores. Sessile marine animals also benefit from protection by secondary metabolites (Green 1977, Hashimoto 1979, Coll et al. 1982, Dyrnda 1986, Hylands 1994). These chemical defenses help mitigate the absence of immune systems in corals as well as in trees.

Concerning Plasticity

Both corals and trees have a high degree of plasticity, a measure of the amplitude of modification that the environment imposes on the expression of the genome (Jennings and Trewavas 1986). If there are instances of morphological plasticity in free-living animals, such examples are unusual. On the whole, animals prefer to flee rather

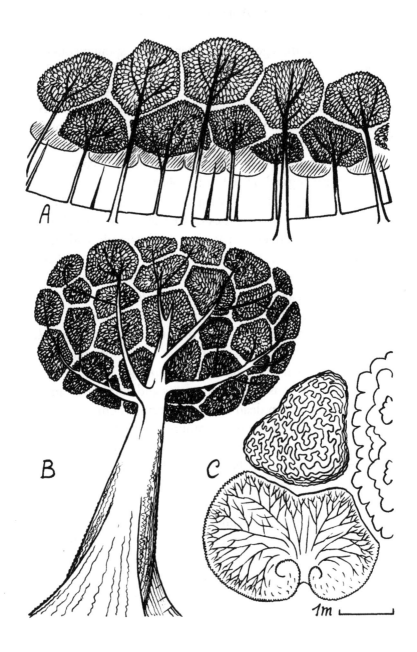

Figure 83. (A) Crown shyness in trees, frequently occurring between trees of the same species, such as the camphor of Borneo, *Dryobalanops aromatica*, or umbrella pine, *Pinus pinea*. (B) It can also be observed between reiterations in a single crown in the meranti, *Shorea*, or the evergreen oak, for example, confirming the colonial nature of these trees when they attain great size. (C) An analogous shyness can be observed between coral colonies of the same or different species (Seribu Islands, Indonesia).

than allow the vicissitudes of the environment to modify their body organization; their behavior, not their bodies, has the greatest plasticity. Recall that the greatest plasticity in plants occurs particularly in those with the longest lives: trees.

Again, corals are like plants in terms of plasticity. Corals have highly developed body plasticity, controlled by depth, currents, the shock of waves, turbidity, the amount of time immersed, etc. This plasticity is so great (Figure 84) that it makes distinguishing between certain species nearly impossible, creating major problems in the taxonomy of corals, of which fewer than 20% of the species can be identified without knowledge of their physical environment (Veron 1995; see also Laborel 1970).

Genetic plasticity is more difficult to evaluate and it raises more serious questions. Recall also that genetic plasticity is not very great in free-living animals, which have genomes stable in both space and time, but that the situation is very strongly the reverse in plants, whose genomes are plastic, even fluid. This plasticity results from at least four mechanisms, common in plants but absent in animals: coalescence of juvenile forms, accumulation of somatic mutations, hybridization between different species, and the development of series of polyploid species.

Although research on genetics has made little progress in corals, we do know that corals possess these four mechanisms and that they have genetic plasticity comparable to that of plants. In the corals *Pocillopora* and *Porites* of the Great Barrier Reef of Australia, colonies are formed through the joining of larvae. Products of sexual reproduction, the larvae are genetically distinct and the adult colony thus becomes a genetic mosaic (Stephenson 1931, Jackson et al. 1985). This same coalescence of juvenile forms also occurs in sponges (Humphreys 1970, van de Vyver and Willenz 1975, Rasmont 1979) and has been discovered more recently in marine algae (Couvens and Hutchings 1983, Martínez and Santelices 1992).

Modified morphological regions, called neoplasms, can appear during development of a coral colony (Figure 85). Somatic mutation is probably the source of such neoplasms, which are comparable to mutant branches in trees (Veron 1995). These neoplasms can be artificially isolated, cultivated, and used to form a mutant colony. The same process is used in fruit tree culture to produce new cultivars from favorable mutant branches.

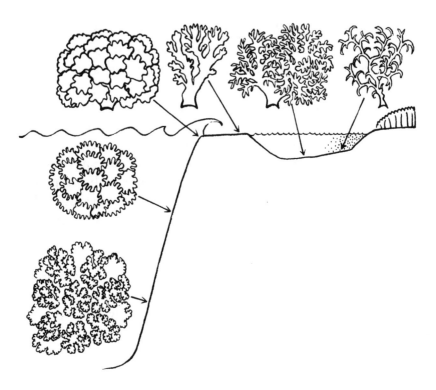

Figure 84. Plasticity in the structure of a coral, *Pocillopora damicornis,* of the Great Barrier Reef, as a function of the environment. From the base of the reef (lower left) to the shallow water of the lagoon, which is marked by turbidity from nearby mangroves, the forms of the colonies vary in a continuous fashion and in such number that they are frequently confused with other species of *Pocillopora* (Veron 1995).

Interspecific hybridization has also been achieved experimentally in corals using the genera *Acropora, Montipora,* and *Platygyra.* Even intergeneric hybrids, *Platygyra* × *Leptoria,* have been obtained. These crosses have not been observed in nature but it is reasonable that they could occur (Veron 1995). Trees and corals seem to function identically in these ways. For both groups, we can hypothesize that interspecific hybridization is a mechanism that increases the genetic plasticity so indispensable for the lives of these immobile beings.

As for polyploidy, recall that it allows a hybrid to have normal pairing of chromosomes during meiosis even if the parents do not

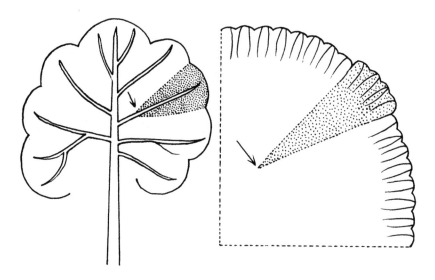

Figure 85. Mutants and neoplasms. Left, a mutant branch in the crown of a tree. Right, a neoplasm in a coral colony. In each, the origin is pinpointed (arrows). These coral neoplasms probably derive from somatic mutations (Veron 1995 and personal communication).

have identical genomes, for example, from different species in an interspecific hybrid. This is why the double mechanism, interspecific hybridization plus polyploidy (Figure 86), is one of the most important in plant evolution, 75% of angiosperms and 95% of ferns being polyploid. A polyploid species series has been discovered in *Acropora* corals (Kenyon 1992). This could be a mechanism for allowing meiotic pairing of different genomes, as in plants. The Australian coral expert J. E. N. Veron (1995) wrote, "The knowledge gained over the last decade has taken corals from the comfortable sphere of understanding traditionally associated with vertebrates and plunged it into the sort of chaos habitually reserved for plants."

Reticulate Evolution

In free-living animals, evolution unfolds along a number of direct lineages that are well marked phylogenetically even if they last only a limited period of time: diversification of cellular function; adaptation to terrestrial life or, on the contrary, specialization for swim-

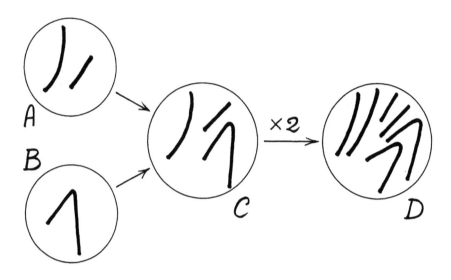

Figure 86. Interspecific hybridization and polyploidy. (A, B) Gametes of two different species whose chromosomes differ in number, size, and structure. (C) The hybrid is sterile; its chromosomes cannot form pairs and successful meiosis is not possible. (D) If the hybrid becomes a polyploid, it is thus capable of normal meiosis and sexual reproduction.

ming; elaboration of the nervous system; etc. In animals, phylogeny can be viewed as directional, illustrated as a tree, culminating in present-day species at the growing tips (Figure 87). The almost total lack of interspecific hybridization preserves the independence of these tips, permitting reconstruction of the distinct ancestral branches.

Neither plants nor corals have such directional phylogeny. We describe their phylogenies as reticulate, marked by absence of direct lineages of any length, by frequency of interspecific hybrids, even by fusion of two species into a single one, forming a network structure (Figure 87).

FINALLY, trees and corals have in common the potential of dominating their landscapes in a manner more profound and durable than can be done by free-living animals.

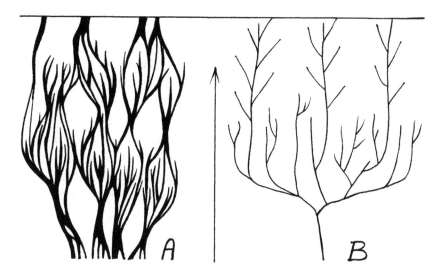

Figure 87. Reticulate and directional phylogenies. (A) Reticulate phylogeny of corals or plants as seen by Veron (1995). (B) Directional animal phylogeny as seen by Gould (1989). The arrow indicates the flow of time. This comparison has no goal other than showing the frequency of fusions in A through interspecific hybridization, and their absence in B. Other phenomena, such as diversification or extinction, are present in both phylogenies.

The Forest and the Reef

Forests and reefs testify to the predominance of life fixed in position in natural landscapes, whether terrestrial or marine. Only sedentary human societies, with their cities and highways, leave an impression on the landscape comparable to that of the Amazonian rain forest or the Great Barrier Reef. What is the basis for this predominance of trees and reefs? Their large dimensions are permitted by their indeterminate growth and unlimited life span, but to those reasons we must add the accumulation of dead tissue. This accumulation is common in both trees and corals but forbidden in free-living animals because it is incompatible with mobility. Harper et al. (1986) remarked that colonial organisms, which they called modular, are capable of accumulating a significant amount of dead tissue, such as the peat that accompanies *Sphagnum* moss, the wood of trees, and the skeleton of coral reefs. The necrotic mass is covered by a thin skin of active and nourishing modules, whether polyps or leaves. This function does not seem to be exclusively excremental (Chapter 4).

Reefs and forests produce canopies, interfaces where biological diversity is exceptionally great. The similarity between trees and corals extends even to the domain of biogeography. In both cases, they predominate in the Tropics (Figure 88), where we find the greatest biodiversity and the largest organisms. The tropical rain forest is more imposing and richer in species than forests at higher latitudes. Coral reefs do not occur outside the Tropics. Lacking symbiotic zooxanthellae and thus incapable of using solar energy, corals there do not form reefs (Hallé 1993).

How to Live Fixed in Place

The tree–reef comparison legitimizes our inquiry into the characteristics of beings who live fixed in place. It also helps us understand better what a plant is in a most fundamental way. Being fixed in place prevents active search for concentrated sources of energy. Such beings must satisfy themselves with uncertain and sparse energy sources, whether luminous or not. They compensate for this handicap by developing vast surfaces for capturing energy. A system of branched axes, growing indeterminately in length and thickness, makes for an optimal arrangement, with radial symmetry permitting energy capture from any solar angle relative to a vertical axis that has a polarity from base to tip.

Indeterminate growth has other advantages. It allows competition with contenders who exploit the same source of energy and the repairing of wounds made by herbivores, who may also be repelled by biochemical defenses.

Among the attributes of those who live fixed in place, we can include architecture, plasticity in form, vegetative reproduction, regeneration of parts, reiteration, and colonialism, adding up to a highly developed mode of life founded on the sacrifice of the individual (Holldobler and Wilson 1994). It would be erroneous to consider that organisms living in place are deprived of sensory organs. Sensibility is not lacking. There is, nonetheless, a sense that they do not have: sight. This is fortunate. We can imagine the frustration of seeing the prey and not being able to draw near, to see the predator and not being able to flee, or even hide!

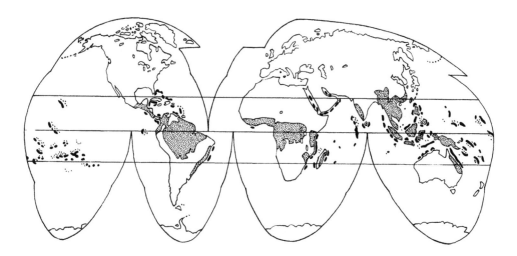

Figure 88. Predominance of organisms that live fixed in place in natural landscapes: tropical forests (gray) and reefs (black). Both find their most complete expression in the Tropics (Hallé 1993).

Being fixed in place prevents active search for sexual partners. The organism must rely on physical vectors or the mobility of other organisms to assure dispersal of its gametes. The latter, along with embryonic forms, represent the only free and mobile stage in the life cycle.

Faced with the hazards presented by living fixed in place, genomic plasticity is an effective remedy, maintained by mechanisms such as coalescence of juvenile forms, somatic mutation, interspecific hybridization, and polyploidy. The lack of separation between germ and soma allows somatic mutations to be transmitted to progeny. All this translates into a reticulated phylogeny.

Finally, organisms living fixed in place often attain enormous sizes and dominate the landscape through the accumulation of dead tissue. They owe their potential immortality to their colonial nature. On the contrary, particular structures are not obligatory to this sedentary life but are the lot of this or that group: Some produce roots, and others, medusas.

I do not claim that all beings who live fixed in place—*Vorticella*, morels, barnacles, lichens, cochineal insects, laminarians, limpets, mussels, *Acetabularia*, chiggers, sponges, remoras—necessarily display all the characteristics enumerated here. Phylogenetically, the sedentary life is more or less ancient, and it imposes a syndrome more or less complete. To compare terrestrial plants with marine animals is a seemingly incongruous enterprise. The evidence provided by this long list of similarities justifies the proceeding *a posteriori*, however, forcing us to defy incongruity and tackle other areas of biology.

Plants and Insect Societies

I dedicate a few paragraphs to a comparison of plants with insect societies. Such comparison is not new; William Morton Wheeler (1911), a Harvard entomology professor, wrote a celebrated article, "The ant-colony as an organism," that goes back almost a century. Since then, the obvious physical analogies between the colony and the organism have continued to fascinate entomologists. Despite highs and lows, these comparisons have not ceased to gain depth and precision (Holldobler and Wilson 1994).

Thanks to Bruno Corbara, an ethologist and specialist in animal social structure, I am able to outline the characteristics of the superorganism—which is what entomologists call it, without specifying plant or animal. The social insects form a world of extraordinary diversity; it is not necessary that a single society exhibit all the characteristics enumerated to be a superorganism. The idea of the superorganism proceeds from a synthetic view of insect societies, introduced by Wheeler and enriched by numerous successors, particularly E. O. Wilson. Ant societies—they are the ones mainly discussed here—require a subterranean portion, the nest, with the queens and their descendants, and an aerial portion, represented by the columns of workers and soldiers charged with capturing prey and bringing them back to the nest. The spatial distribution of these columns resembles the filaments of fungi growing at the surface of a nutritive medium (Figure 89). If the food is distributed evenly, fungal hyphae and insect columns adopt the same distribution, a regular star around the point of departure (Rayner and Franks 1987). On the contrary, if nutrition is locally concentrated and unequally distributed, a system of marking must be established, and the ant columns take on the appearance of trees, with trunks and branches.

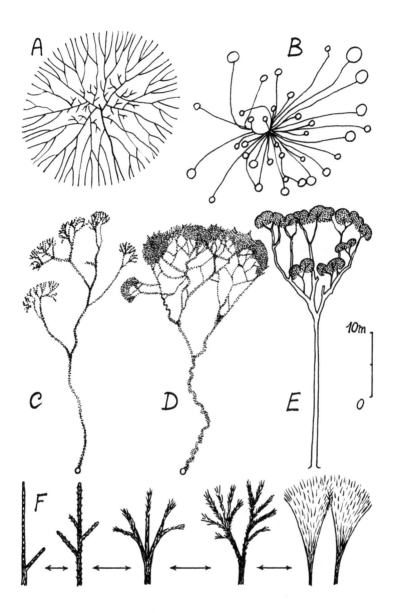

Figure 89. Ant societies and their analogues in other organisms. When food is distributed homogeneously, fungal hyphae (A) and ant columns (B) adopt the same star-like distribution around the point of origin. If food is unequally distributed, the columns of ants take on the appearance of trees: columns of the ants *Eciton rapax* (C) and *E. burchelli* (D; Holldobler and Wilson 1994). (E) For comparison, the gonfolo tree, *Qualea rosea* (Vochysiaceae) of the Guianas and Brazil. The scale is the same for B–E. (F) For ants as well as plants, the search for energy leads to the adoption of axes more or less branched; apical dominance is stronger or weaker (Rayner and Franks 1987).

The search for localized sources of faraway food leads ants, like plants, to adopt a mechanism of collective coordination, which is translated into the formation of columns pushing outward at their tips. As a function of the degree of strength of this coordination, an apical dominance is established more or less strongly, and columns, more or less branched, can be observed. Disappearance of apical dominance leads to the appearance of fans: Growth slows and exploration loses its place to exploitation, according to a typical plant strategy (Figure 89). "The societies of most ant species . . . reproduce like plants. They throw out large numbers of colonizing queens, like so many seeds, on the chance that at least one or two will take root" (Holldobler and Wilson 1994).

Decentralized operation helps ant societies resist the vagaries of climate. Nest temperature is maintained at an optimum, about 30°C, and relative humidity is higher inside than outside. Climate control is even more effective in termite nests since those insects establish true air-conditioning. In addition, termite mounds, often of elaborate form, are repaired after falling down according to a process that evokes the reiteration of a fallen tree (compare Figure 90 with Figure 60, Chapter 5).

Decentralized operation also makes possible the budding that characterizes polygynous species (in which each society includes several fecund queens). It is experimentally possible to remove a portion of the society and implant it at a different site. The missing queens are replaced and a complete society regenerates. Some societies are capable of reproducing themselves spontaneously by budding in a manner of a tree that suckers from its roots. Since societies produced in this way do not separate from the initial one, a federation of interconnected nests can attain enormous size. A federation of *Formica yessensis*, observed on the coast of Hokkaidō in northern Japan, was composed of 45,000 interconnected nests, comprising hundreds of millions of workers and queens. The absence of aggression between ants from different nests indicates the territorial extent of the federation. This zoological equivalent of a plant clone covered 270 hectares and, like the plant clone, is potentially immortal (Holldobler and Wilson 1994). To die, said André Comte-Sponville, is the price paid to be one's self.

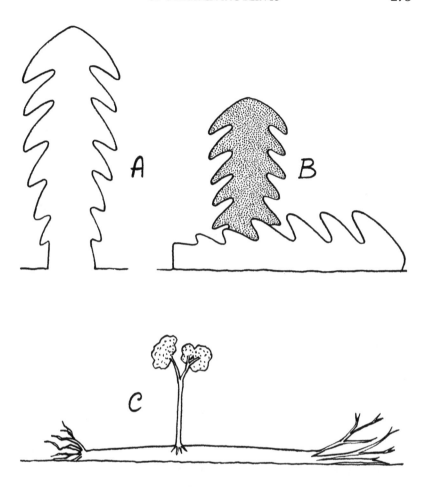

Figure 90. Reiteration in a termite mound. (A) The original mound. (B) After falling, the mound is repaired by a process of reiteration (Republic of the Congo; Hubert de Foresta, personal communication). (C) The process appears identical to that observed in fallen trees (Negrelle 1995).

Evolution of Behavior or of Form?

I borrow the thread of the following from Bruno Corbara, who will recognize his research question here. Is one of the most important differences between a plant composed of cells and a nest composed of ants that the constituent elements remain aggregated in the former and dispersed in the latter? In both cases, these elements must exchange enough information that cohesion of the assemblage is

assured. The olfactory messages that ants share have their counter-
parts in the chemical messages that unite the cells of a plant during
growth. In both cases, a division of labor is imposed that leads to
differentiation. Evolution that confers adaptive behavior on insects
for each function, leading to differentiation into castes, finds a parallel in evolution of form, in which cells differentiate so that all necessary functions will be fulfilled, to the benefit of the entire plant
with its various organs.

Fixity in place, resistance to change, morphogenesis and reiteration, regeneration and budding, unlimited growth, and potential
immortality conferred by colonial structure—these are characteristics of insect societies that evoke plants. I leave to the entomologists
the task of determining if their superorganism is a reality or spiritual
view. If it exists, it is certainly a plant.

Looking for Analogues

Where must we look to find other forms analogous to plants? Beyond the limits of life, without doubt. Several vegetable analogues
have been found outside the living world. When manganese oxide
infiltrates a fissured rock, we see plant-like designs that geologists
call dendrites (Figure 91D), a term evoking plants (*dendron* is Greek
for tree). A layer of horizontal branches in *Araucaria* merits comparison with a snowflake (Figure 91B, C).

Impassioned by ferns since infancy, the mathematician Michael
Barnsley formulated a hypothesis that a fern spore contains only a
limited amount of information. If this information expresses itself
morphologically by iteration, the fern itself is obtained. Starting
from a frond of *Asplenium*, Barnsley (1985) constructed a fractal
image by the iteration of several simple morphological functions
(Figure 91A).

It is already an ancient scientific tradition in France to compare
plants with crystals. Auguste Bravais, one of the founders of crystallography, was first interested in plants. His father led him on botanical excursions around Annonay, his village of birth. All his life he
used plant patterns in his work. In Algeria, he studied the precise
network of areoles that mark the stem of the so-called fig of Bar-
bary (*Opuntia ficus-indica*, a cactus). It is reasonable to conclude that
he owed his first intuitions into the physics of solids to this plant. In

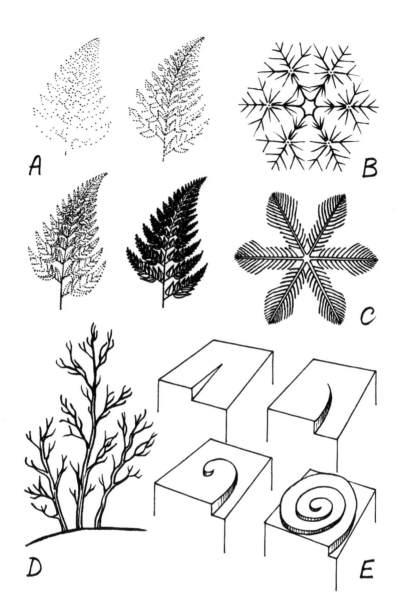

Figure 91. Some analogues of plants found outside the living world.
(A) A fractal image approximating the frond of a fern, obtained by iteration
of simple mathematical functions (Barnsley 1985). (B) A snowflake, for
comparison with (C) a single level of horizontal branches of a tree, *Araucaria*,
viewed from above or below. (D) Dendrites formed by the infiltration of
manganese oxide into a rock fissure. (E) Growth of a crystal at the molecular
level. A screw dislocation attracts molecules from the surrounding solution,
making the crystal grow in length (Kittel 1958), an analogy relevant to the
action of a meristem constructing a branch.

1854, received into the Academy of Sciences for his crystallographic research, he described his intellectual trajectory: "First, there are my botanical studies on the spiral arrangement of leaves in plants, which led me to consider the geometry of the network of points. It is this latter study, along with the notion of the mode of coordination of crystal molecules, that inspired me to write my article on the assemblies of points in space."

Bravais inspired successors, among whom some saw interesting resemblances between plants and crystals. For Henri Doffin (1959), plants *are* crystals, and his arguments merit some reflection. A crystal, like a plant, is the result of growth. In a saturated solution following appropriate seeding, a crystal grows by fixing atoms or molecules on its surface where the concentration of the solution is lower, near the growing face of the crystal, before the establishment of currents that tend to maintain concentrations in equilibrium. A steady state is established, the supply of solute from more distant regions of the solution balanced by crystallization. If molecules are replaced by cells, the ensemble begins to function like a meristem. For Doffin, a plant is an elongated crystal, the latter termed a whisker, in which the cell is the constituent unit.

Crystalline growth relies on a screw dislocation, an unevenness running across part of the face of the crystal (Figure 91E). Molecules coming from the ambient solution to the crystal lodge along this dislocation, which tends to propagate itself. Because one of the extremities is fixed in place, crystallization occurs spirally (Kittel 1958). We find as many left-handed as right-handed spirals.

Is a Plant a Crystal?

For Doffin, crystalline growth made the fundamental spiral of plants comprehensible, as well as the equivalence in number of left- or right-handed plants (Figure 92D), just as in enantiomorphic crystals. In plants, the spiral carries leaves, the phyllotactic spiral visible to the naked eye on the stem of any plant with alternate leaves—pine, oak, or sunflower (Figure 92C). Interactions between two spirals of opposing handedness make the arrangement of opposite or whorled leaves understandable (Figure 92A).

The screw dislocation being an unevenness in the crystal, its presence at one end imposes that of another screw dislocation at the

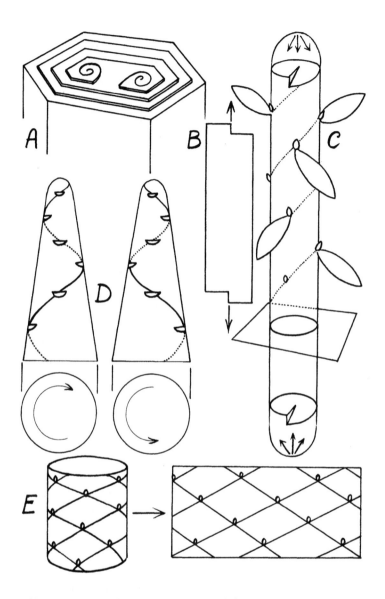

Figure 92. Comparison of plants and crystals. (A) In crystalline growth, interaction between two opposing screw dislocations leads to the formation of terraces, analogous to a whorled phyllotaxis (Dekeyser and Amelinckx 1955). (B) Shoot and root together are comparable to a crystal growing at both ends. (C) A plant can be thought of as a growing crystal in which the fundamental element is the cell and on which screw dislocations are lateral buds (Doffin 1959). (D) In plants as in corals, the two directions of spiral growth occur in equal frequency. These two plants, one the mirror image of the other, are enantiomorphs. (E) The phyllotactic network, hardly visible on a cylindrical branch, becomes obvious if the cylinder is unrolled.

other end of the crystal; the system grows at its two extremities, like the ensemble of a shoot and root (Figure 92B). As in plants, there is no theoretical limit to the growth of a crystal; a larger one can always be obtained if the proper conditions exist.

The periodic appearance of leaves along the spiral could be a manifestation of what crystallographers call epitaxy (Joël Doffin, personal communication). If a substance begins to crystallize according to some given system, before being replaced by a second substance whose matrix of crystallization is a little different, we observe the adoption of the first system, followed by the periodic appearance of screw dislocations that constitute points of lateral growth (Figure 92C). In plants, the latter correspond to axillary buds situated at the nodes of the stem. The different substances in the crystallographic system could correspond, in plants, to different strata of cells produced by the apical meristem (Figure 62, Chapter 5). We can conceive of plants as crystalline in growth where the basic element is the cell. This concept is not immune from criticism. For instance, we can object that the forces assuring cohesion between cells has nothing to do with those that unite atoms in a crystal. It remains, however, that in crystals as well as plants, forces of cohesion exist between the constituent elements.

In a celebrated book, Jacques Monod (1970) argued for a crystalline model of life. Unlike objects, natural or artificial, that owe the essentials of their form to the action of external agents, crystals, like living beings, are distinguished by the autonomy of their morphogenesis. The visible structure of a crystal directly reflects the spatial arrangement of the atoms that compose it, a little like how a genome is reflected in the form of a living being. The crystal, Monod wrote, "is the macroscopic expression of a microscopic structure." Also, in both cases a germ is necessary to initiate morphogenesis, and "by the quality of uniform reproduction, living beings and crystalline structures find themselves at once the most similar and different of all the objects in the universe." This idea attracted a small following (Cairns-Smith 1971) but I end the plant–crystal comparison here, hoping that my colleagues, the physicists, excuse the crystallography described here by a botanist, simple as it is.

Stéphane Douady and Yves Couder, physicists at ENS (École Normale Supérieur), Paris, experimentally constructed a plant analogue whose degree of realism is particularly striking (Douady and

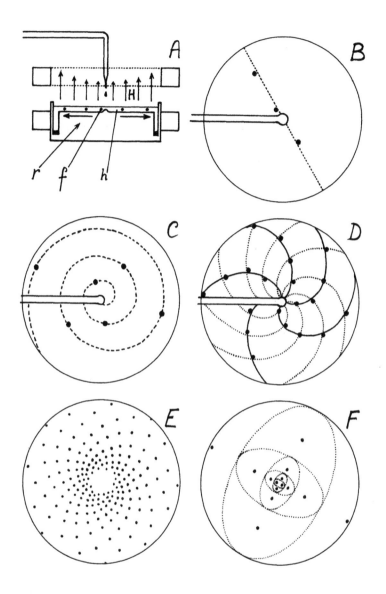

Figure 93. Physical model analogous to a meristem. (A) The experimental apparatus consists of a rotating disk (r) covered with a layer of oil (h), at the center of which drops of a magnetic fluid fall. A magnetic field (H) transforms the drops (f) into small magnets that repel each other. By varying the periodicity of drops and the force of the magnetic field, a distichous phyllotaxis (B) or a phyllotactic spiral with an index of $^2/_5$ (C) may be obtained. (D) Weak repulsion between drops corresponds to a parastichous phyllotaxis of higher order. Numerical simulations permit replication of very rich phyllotaxes, such as a sunflower head (E) or an opposite and decussate (F) leaf arrangement (Douady and Couder 1992).

Couder 1992, 1993). Calibrated drops of a magnetic fluid are made to fall at the center of a rotating disk covered with a layer of oil. A vertical magnetic field sweeps these drops toward the periphery of the disk, changing them into small magnets that repel each other (Figure 93). This force of repulsion directs each drop toward the zone of least resistance. The disk and the magnets are analogues of the apical meristem and the leaf primordia formed by the meristem, respectively. A simple change in the timing of the drops—what botanists call the plastochron—allows the experimenters to simulate different leaf arrangements (phyllotaxes), distichous or spiral, with the fewest rows of leaves (called parastichies) occurring when the force of repulsion between drops is greatest. A numerical simulation founded on the same physical hypothesis permits replication of all the known phyllotactic arrangements. Douady and Couder inferred that control of meristematic form and function is almost exclusively external, with the role of genes limited to the specialization of organs.

Immanence and Transcendence

An overall view of plant analogues, products of a nonliving world, can lead us to contradictory positions. Given that the constituent elements and the forces that assemble the living versus the nonliving are so different, it seems legitimate to consider resemblance as chance, and superficial. At present this is the common opinion, or, in the research environment we tend to look on these analogies with amusement, occasionally influenced by aesthetics and curiosity.

We can also adopt a more positive point of view. If it is impossible to evaluate the conclusions that can be extracted from the study of these analogies in detail, can they not shed light on our understanding of plants, their constitution in the physical world, their conformity to ecological rules, their adaptation to the network of constraints dictated by living a life fixed in place?

Certainly, animals must accommodate themselves to physical laws, but in ways appropriate for them. From the fact that they are mobile, they flee from difficulties rather than submit to them. Perhaps animal evolution can be presented as an increase in the number of degrees of freedom, leading to acquisition of more and more complete autonomy from the constraints of their environment (Favre and Favre 1991). On the contrary, plant evolution would be adaptation more and more driven by hard reality, with more and more complete integration into the environment (Rose Hébant, personal communication). Animals present images of their own freedom; plants, more modest, express the constraints of the environment. Each conveys the contrast in its own way. For Aristotle, the animal was a "hungry spirit," and the plant, a "diffuse vegetative spirit" (Barnes 1984).

> Respect an active spirit in the animal . . .
> Each flower is a spirit of Nature born.
> GÉRARD DE NERVAL, *Les Vers Dorés*, 1922

Francis Ponge (1942) spoke of trees: "They are only an expression of will. They have nothing to hide for themselves; they guard no secret idea, they arrange themselves entirely and honestly without restriction . . . , they only occupy themselves to accomplish their expression: they prepare, they decorate themselves, they wait for someone to come and read them." René Thom (1988) was scarcely more prosaic: "One fundamental constraint for the dynamic animal, which distinguishes animals from plants, is predation. . . . Plants do not have specialized prey, thus they always seek to identify with a three-dimensional milieu." For plants, "We find a sort of fractal dilution in the nourishing ambient milieu." Perhaps the transcendence of animals, including humans, is mirrored by the immanence of plants.

CHAPTER 7

Ecology

Fauna that inhabit my paternal earth,
Who lead our dances and our visits to the Loire,
Befriend the plants and give them aid,
That summer does not burn, and winter does not freeze.
Pierre de Ronsard, Sonnets pour Hélène, 1578

The word for world is forest
Ursula Le Guin, 1972

And that we stretch out an arm
Over the sea, toward Brazil,
To harvest fruit from the islands
Encompassing all of the Earth.
Jules Supervielle, Gravitations, 1925

And all the while green tamarinds exhale
Perfumes that fill my nostrils and my soul
Blending with sounds of sailors' barcarole.
Charles Baudelaire, Les Fleurs du Mal, 1861

To walk in a forest between two rows of ferns transformed by
the autumn, that is a triumph. What is that next to approbation
and ovation?
Émile Michel Cioran, De l'Inconvénient d'Être Né, 1973

I RETURN to the fact that animals and plants are different as biological entities. Animals are generally individuals; plants, generally colonies. This confirms Harper's (1977) point, of "the most
striking analogies between the models of interaction between animal
populations and models of plant interaction, as might be expected
from the fact that the animal has the potential for exponential increase in numbers and the plant has the potential for exponential

increase in size." A plant is equivalent to an animal population; a plant is a metapopulation (White 1979). There lies an important point in ecology, and it is not certain that all its consequences have been explored.

Give Plants Their Due

One of the axioms of ecology is that the success of a living being is measured by the number of descendants it leaves; that would be true for plants as well as animals (Hendrix 1988, Waller 1988), and success (fitness) in the two kingdoms should thus be reproductive. This is a valuable concept in animals but when applied to plants it loses sight of the fact that for many, the potential for immortality renders sexuality largely inoperative (Bradshaw 1972). Each year my oak produces a good 1000 acorns; in the course of its life it will have been able to produce at least 100,000 acorns. Each spring the ground is covered with seedlings that quickly die. This may not translate as a check on reproduction because one seedling surviving each century is sufficient to replace the oak. As for those that die, they are not lost for they are decomposed by microbes and return to the soil where the oak can reuse their constituents.

Another example: The theory of population genetics considers plants and animals as entities at the same level without taking into account that the plant, if it has the structure of a population, can contain several genomes. To attribute a single genome to a perennial plant as is to account for only a portion of the variability of the population that in reality is situated within the plant (Murawski 1998). Who could doubt that what gives plants their deep biological significance could also advance ecological understanding?

Another way of gaining greater ecological insight is to compare the feeding habits in the two kingdoms. Apart from differences in energy capture (Chapter 2), a profound but rarely mentioned divergence exists between the modes of feeding in plants and animals. Animals have alimentary habits of extreme kinds. In addition to the voracious ones—sharks, rats, seagulls, cockroaches, humans—capable of gulping down just about anything, there are animals that avail themselves of distinctive food sources: marine plankton, garden soil, pollen, blood, sweat, rotten wood, mildew, spoiled cheese, vaginal secretions, or poorly maintained zoological collections. Others, more refined, have a particular food on which they feed exclusively: rose sap, octopus urine, passionflower foliage, paralyzed spiders, pea cotyledons, blood of swifts, or bat dung. That these menus are not very inviting is another matter; I only wish to say that where there is food, an animal will be adapted to eat it.

What is interesting, here again, is a comparison with plants. The essential nutrition of plants consists of ubiquitous substances: carbon dioxide, water, mineral salts. The latter vary depending on the soil, which can scarcely be chosen by the plant. As for solar energy, which allows use of all these trivial substances, it is also required by plants. Plants are not in a position to choose their food, and in fact, they all obtain nourishment the same way, or nearly so. The differences are only quantitative: a little water for a cactus and lots for a water lily, weak illumination for an African violet and strong for a date palm.

Looking more closely, we can say that reality is not quite so simple because there are plants whose nutrition is specialized: plants of the shaded understory such as *Geogenanthus* (Commelinaceae) of the Amazon, parasites such as mistletoe, saprophytes such as the orchid *Neottia*, or carnivores such as *Pinguicula*. However, these particular cases bear witness, once again, to the versatility of plants. Nevertheless, nutrition is practically the same for all plants—the same menu for a vanilla and a primula, for a baobab and a daisy.

Nutrition and Biological Types

Is it of any consequence that some must content themselves with ordinary soup while others have access to refined, exclusive tidbits? Here it is necessary to consider the biological types of plants, an idea that was popular in France to the end of the 1950s, forgotten, then revived under a name, plant functional types, which means the same thing. Any group of plants tends to diversify into a range of biological types. Take the family of gentians, Gentianaceae. They are widely distributed, especially in alpine tundra, and the stemless gentian *(Gentiana acaulis)* is one of the marvels of the flora of the Alps at the beginning of summer. This small plant, with such large, dark blue flowers, reappears each year in the same place. It is thus a tenacious perennial herb (Figure 94A).

The snow gentian *(Gentiana nivalis)* appears later in summer, taller than *G. acaulis* and with flowers of the same color but smaller

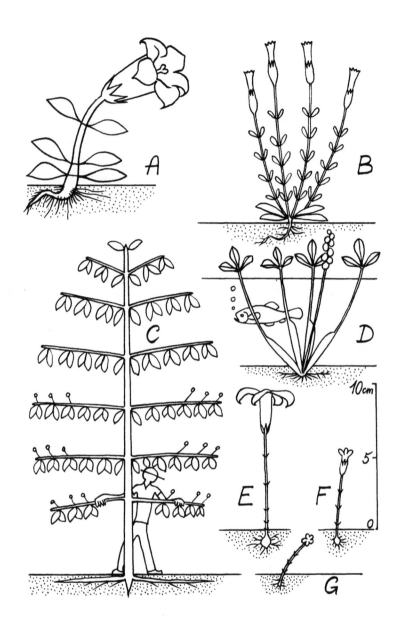

Figure 94. Biological types. (A) Stemless alpine gentian, *Gentiana acaulis*, a perennial herb. (B) Snow gentian, *G. nivalis*, an annual herb. (C) *Tachia guianensis*, small tree of the Amazonian forest with flowers on the lower branches. (D) Bog bean, *Menyanthes trifoliata*, a hydrophyte or aquatic plant attaining a height of 1 m. (E) *Voyria rosea*, a saprophyte from the Guianas and the Amazon. (F) Another saprophyte, *Leiphamos aphylla*, from the same region. (G) *Sebaea oligantha*, a saprophyte of the forests of the Congo Basin. The scale is for A, B, E–G.

and more numerous. It disappears in the fall, passing the winter as seeds in the soil. It is thus an annual herb (Figure 94B).

In the river marshes of Brittany or the Charente, it is easy to encounter the pale flowers of *Menyanthes* and *Nymphoides* (Menyan-thaceae, sometimes included in the Gentianaceae) sticking out of the water. Even though these are also perennial herbs, their aquatic habitat has led botanists to call them hydrophytes (Figure 94D).

In South American tropical forests, in the Guianas and Peru where the understory is a bit clearer, a small, inconspicuous tree is encountered with handsome white or rose flowers, depending on the species. This is *Tachia* (Figure 94C), also of the gentian family. In the same Guianan or Amazonian forest, in sites where the understory is deeply shaded, delicate blue, rose, and white flowers are seen sticking out of the leaf litter, deliciously scented, recalling those gentians of the Alps a little. Their minuscule leaves, gray and without chlorophyll, are incapable of photosynthesis. They receive their nutrition from symbiotic fungi that lodge in their subterranean organs. These astounding plants, belonging to the genera *Voyria* and *Leiphamos* (Gentianaceae), are saprophytes (Figure 94E–G).

The small family Gentianaceae, with 900 species, has developed five biological types. Certain very large families, such as the Euphorbiaceae (more than 5000 species) or Rubiaceae (7000 species), produce many biological types: There are xerophytes that have the appearance of cacti in dry regions, whereas in tropical rain forests we find representatives in the form of trees, lianas, and epiphytes. Certain families contain only a single biological type, such as the Fagaceae in which all are trees: oak, *Lithocarpus*, beech, chestnut. However, such families are rare; most are diversified into different biological types. In contrast, animals are diversified in their feeding behavior. This point of view finds agreement among most biologists (Andrews 1991) despite the fact that it leads to a comparison of functions that are not directly comparable.

Are there equivalents in animals to biological types in plants? We could be tempted to respond affirmatively, grouping bats and owls, crickets and moles, even sharks, ichthyosaurs, sea lions, and dolphins, together in the same biological types, but we quickly learn that this leads nowhere. A dolphin resembles a shark but this is the result of convergence, an analogy emphasizing the fact that the animals share the same environment and are similarly equipped for rapid swimming, which allows them to assume the same role: predator. Apart from that, their structures are different both anatomically and physiologically; they have little in common because they are not of the same evolutionary lineage, and their biology and behavior are profoundly different. That one is a fish and the other a mammal separates them despite their similarity as quick carnivorous swimmers.

When two plants are of the same biological type, an evergreen oak and an Aleppo pine, for example, the resemblance goes much deeper. Even if they are not of the same evolutionary lineage, the two trees make do with the same energy, the same water, the same mineral nutrition. They also have the same structure: branches, roots, leaves, meristems and cambium, wood, etc. This resemblance reaches the level of a homology. A Mexican cactus and a cactiform Somalian euphorb are plant homologues. François Jacob (1970) wrote, "Homology describes the correspondence of structures, analogy that of functions."

Research into biological types in animals seems to be an enterprise doomed to failure. It would show few resemblances beyond those that animals preserve as evolutionarily acquired and unchangeable vestiges that remain in spite of adaptation. Plants are more completely immersed in their immediate modes of life, not preserving much from their ancestors. They play their roles more completely than animals, no doubt because they are fixed in place whereas animals move. We must admit that these aspects are poorly understood.

Ambiguity in the Relationship Between Eater and the Eaten

In the discipline of ecology, the complementarity of plants and animals is a remarkable fact—so remarkable that it is known to all, and unnecessary to publicize. I am satisfied with summarizing the essentials. The energy used by multicellular life is, with very few exceptions, furnished by the sun. Some benthic marine animals—worms, mussels, clams—can survive on energy that does not come from the sun since they live in the depths of the oceans. They live symbiotically with bacteria capable of utilizing energy that is not solar, but chemical (Corliss et al. 1979). No plants are necessary, animals alone being the primary producers; these are "animal-plants" (Françoise Gaill, personal communication).

Plants are capable of feeding directly on solar energy with a contribution from some trivial and ubiquitous materials—carbon dioxide, water, soil minerals—to make their own structures. Good examples of self-sufficiency, they need nothing they cannot find everywhere, and it is for this reason they qualify as autotrophs. This confers on them a sort of magic to which I have been sensitive since my first contact with them. Asking nothing from anyone, collecting and concentrating a form of energy available to all, they produce living material without respite. If they assume the role of producer, it is not only for them, for they also work for all other organisms. Notably for us, plants are the only simple means at our disposal for concentrating energy, which is the origin of agriculture.

Animals are incapable of making direct nutritional use of solar energy; they use it to warm themselves but not for food. Thus to survive they must make use of this energy secondhand; they are said to be heterotrophs. They procure energy by eating plants if they are herbivores—saupe (an algae-eating Mediterranean fish), aphid, lamb, hoatzin (the South American bird *Opisthocomus hoazin*, which feeds mainly on plants of the aroid family, Araceae)—or by eating herbivores if they are carnivores—sardine, owl, louse, vampire bat, ladybug. Half of all insects and 65% of mammals are directly nourished by plants (Hendrix 1988). Just like the customer at a supermarket, the animal is a consumer and disperser of energy; that does not diminish the affection I have for them, but it is good to say these things from time to time.

A direct consequence of autotrophy versus heterotrophy is that plants and animals occupy very different positions in what ecologists call food chains (Figure 95), traditionally presented with plants at the bottom and animals on top. This somewhat triumphal ascendance masks the autonomy at the base and the dependency at the summit. It must be said clearly: One is the eaten, and the other, the eater. To be objective about the relationship between the one and the other is not easy, for we are on the receiving end. Does the animal not eat the plant? Very well, no objection. As a chauvinistic animal, we see a sort of implicit victory. We can eat this plant without its being able to defend itself, without even protesting. That confirms the superiority of animals, also that the plant is made to be eaten. *Vae victis!*

A longer view would show us that in reality, for animals (including humans) there is an implication of, if not defeat, at least a hand-

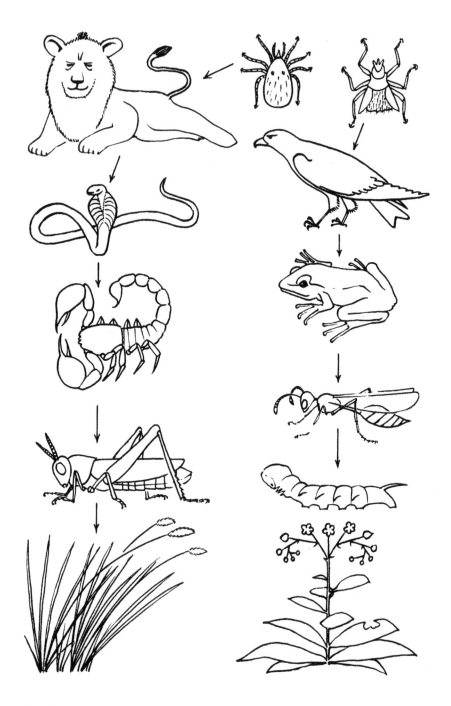

Figure 95. Food chains, two examples. Parasites are at the top, plants at the base.

icap in this complete dependence on plants: Plants have no need of us, yet we need them. So, therefore, our feeding relationship with plants is ambiguous because we only see the part of it that benefits us while the true conquerors do not speak. It is in the nature of the yang to believe itself superior to the yin.

Animals depend on plants in a number of ways, not only for respiration. After having made our terrestrial atmosphere breathable, plants continue to regenerate our air by removing carbon dioxide and supplying oxygen. When tropical forests burn, as in the Amazon or Indonesia, we have a hand in destroying the factories of purification that trees are, especially if the fire is accompanied by a return of toxic compounds to the atmosphere, compounds accumulated with the patience and discretion that characterize trees.

The atmosphere is not our only interest. Plants are sufficiently numerous and vast to create the microclimates that most animals appreciate. These microclimates include the coolness of an oasis, the calm of a garden surrounded by a windbreak of fruit trees, the ambience of an esplanade under rows of plane trees, the humid understory of a beech grove, or an Andalusian patio with walls covered in geraniums, bougainvilleas, and monsteras, accompanied by the sound of a trickle of water, in the heavy heat of summer.

Climates and Landscapes

No free-living individual animal creates a landscape by itself. Social animals can mark their environs, whether it is an island of guano, a savanna covered with termite mounds, a reef, or even a town. Plants, through their size and longevity, collectively establish the landscapes that provide the framework of life for most constituents of a fauna, including humans (Figure 96).

Certainly, geology plays a major role in the genesis of landscapes, especially through tectonic movement. Plants arrived soon after rocks, helping form soils as much as through the physical effects of root growth as through chemistry and supplying organic material. With the exception of reefs, designations for biotic landscapes are botanical: garrigue, matorral, varzea, igapo, bush, chaparral, thicket, caatinga, cerrado, steppe, mangrove, savanna, woodland, lawn, prairie, forest. No mention of fauna—only references to flora designate these natural landscapes (Darley 1990).

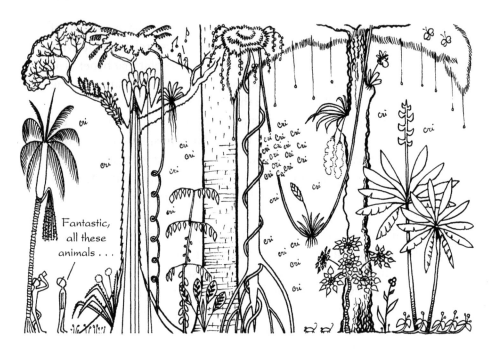

Figure 96. Plants make the landscape. Animals, including humans, use these landscapes as a framework for their lives.

Influence on general climate, as well as on microclimates as mentioned, is another difference between plants and animals. A single large tree may have 160 hectares of moisture-exchange surface area (Chapter 2), almost the size of the principality of Monaco. The formidable surface areas of vegetation promote humidification of the atmosphere through evaporation and transpiration. Plants clearly participate in the formation of clouds. Do they also affect rainfall patterns? It would seem that the answer is yes, but there is less certainty on this subject. Under the pure sky of the savanna, forested areas are discernable from afar by the clouds that rise above them, just like islands on the ocean horizon. After a long walk in the Tropics, leaving the burning sunlight for the freshness of the forest margin, I often encounter rain, an agreeable contrast that reminds me of Brittany.

Those who are well acquainted with the forest know empirically that its presence ensures a regime of moderate rains, regular and lasting. They often note that deforestation at a given site is exchanged for bursts of brief, brutal, devastating rains. These are accentuated year to year by the disappearance of forest (Charles Baldy, personal communication). It is reasonable that deforestation in a region leads to reduction in annual precipitation but this has not been directly demonstrated. It is necessary to recognize that it is a difficult hypothesis to test since long-term meteorological records are generally not available for these forests; record keeping commences after the forest has disappeared.

In contrast, and no one doubts this, forests regularize runoff and influence streams and rivers. Actual conditions in many mountainous areas—in Colombia, the Philippines, the Himalaya—provide evidence that deforestation increases flooding in the low valleys. No animal community can provide protection comparable to that of the forest, and I regret to add that if animals do not create landscapes, they are capable of destroying them. Look at a field of vegetables under clouds of migratory locusts, or a banana patch after the passage of elephants. This leads us back to the subject of eating, where the dependence of animals on plants is most obvious.

Who Needs the Other Most?

Remove the fauna from a forest, leaving only the plants. What will happen? Nothing—for years, perhaps even centuries. Plants will continue to grow normally. The flowers of some will not be pollinated as efficiently, but wind and mechanisms promoting self-fertilization will at least partially mitigate the absence of pollinators—bees, sphinx moths, hummingbirds, bats. Seeds will be produced in smaller number, it is true, but is this so serious? Animals are not there to levy their habitually enormous tax, which is a major cause of plant mortality.

Animals—squirrels, ants, elephants—will no longer be there to disperse seeds, which will tend to germinate close to the trees that produced them. It is reasonable to expect, after a delay of several centuries, a slow change in the distribution of a few species in the forest (Daniel Sabatier, personal communication). It is possible that some tree species will eventually disappear, those particularly dependent on animals for the unfolding of their sexuality, such as figs, trees with large and few pollen grains, or trees with separate sexes: nutmeg, pistachio, *Raphia* palm, elm, laurel. Still, that is not certain; plants can be aided by two mechanisms capable of saving them, reasonable alternatives:

> Increased vegetative vigor, a normal consequence of the loss of sexuality or capacity of a plant to produce descendants (Figure 97)
>
> Vegetative multiplication through layering, suckering, etc. Sterile plants such as water hyacinth *(Eichhornia crassipes)* or lalang grass *(Imperata cylindrica)* hold records for efficiency in vegetative reproduction; the absence of sexual fertility does not necessarily make plants disappear and seems to have made some capable of invading entire continents

We thus should not be too concerned about our trees being deprived of descendents; they will get by. Plants have a long history of countering difficulty. Unlike animals, they do not flee, turning challenges to their advantage through a combination of internal genetic variability and longevity. Bacteria and fungi, not being animals, are present and continue to live off the organic material produced by the plants. Neither recycling nor root–microbe symbiosis is affected.

It seems, without being able to establish incontrovertible proof, that the forest will continue to live, and live well, indefinitely. I can even imagine it slowly regaining lost territory and covering the continents again, as it had before the arrival of its single true enemy, the human being. I have a conviction that things would pass thusly for the forests in temperate regions, where animals do not play decisive roles in the biology of plants. For tropical forests, I think it might be similar, prudently leaving the question open.

Now remove all the plants from a forest, leaving only the fauna. There is no doubt about the outcome. Wait several days, even several hours, for a nightmare. At first, carnivores will continue to feed on the herbivores, and when the latter begin to die of hunger, the carrion eaters will rip apart the first cadavers. Wait for the howls of panic, for the clusters of ants gnawing at the feet of mammals, for the swarming of maggots in the carrion when, at the conclusion of this hecatomb, the horrible odor of death rises, hideously and quickly. Bacteria and fungi, not being plants, are present, studiously taking care to clean up all that can be in good time.

We humans, so imbued with our superiority over the ensemble of living creatures, do we believe we could survive the disappearance of the world's plants? How much time would we have? Several weeks,

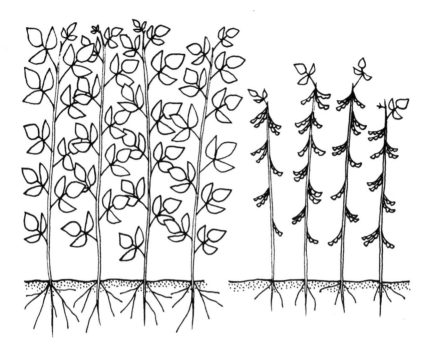

Figure 97. The negative effect of fruiting on plant growth. These two lots of soybean are the same age and have been grown in the same environment. Left, fruiting was prevented by removing flowers to determine what the plants would otherwise produce during the growing season. Right, the control lot, with normal fruiting (Leopold 1978).

or rather, several days? We should not fool ourselves—without plants, humans would rapidly disappear, like all animals. These things must be said, and our vanity must suffer accordingly. Humans have a power over other living beings that we, ourselves, judge as excessive. While waiting for a counterpower to manifest itself, we should not lose sight that humanity, despite the technical prowess of which we are so proud, is completely incapable of fabricating the sprig of a plant or the stylet of an aphid.

All animals disappear once deprived of plants. Almost all, I should say, because certain animal species would do better than us. Those at the bottom of the ocean, thanks to bacteria, know how to use energy other than sunlight. Suppress the plants and soon there would be no animals except for some obscure mollusks and ungainly worms, grouped around those fumaroles in the abyss that are called black smokers.

Noah's Two Arks

I have it in my heart to right an injustice: The vision of a fauna deprived of a flora has a rather troublesome historical precedent. In the classical version of Noah's ark, the story would end very badly for the animals for ecological reasons that escape no one's notice. How was Noah, that ungrateful person, able to forget the grapevine? Not one, but two Noah's arks are needed (Figure 98). The first would be called *Phyton* and would be the flagship. The other, *Zoon*, would be of little interest and left behind.

Figure 98. Noah's two arks, in the style of Dominique Appia's (1977) *Grand Voyage.*

Epilogue

Because he is and understands everything from his own experiences, because a certain similarity with his behavior is the *sine qua non* for him to notice traces of intelligence outside himself, one can understand how difficult it is for him to imagine emotions without a tangible substrate, where he usually encounters them.

MICHEL LUNEAU, *Paroles d'Arbre*, 1994

To spiritual eyes the plant is no longer a simple object of humble and passive life, but a strange prayer of the universal warp. (*"trame universelle"*)

PAUL VALÉRY, *Dialogue de l'Arbre*, 1943

THE SPIRIT OF PLANTS.—Plants have their shadow that is the spirit itself, that is why they need to be touched so that the spirit feels and moves . . . and they also have eyes. More eyes than leaves, for each leaf they have two and they are the same as ours but tiny. Of course it is beyond our reach to discover them, but they are there, for the plants to see.

DON SANTIAGO CAICEDO, *El Mundo y Sus Cosas*, 1993

Surely, I love flowers. But I love animals just the same; without a doubt the anemone knows that, since its heart in full bloom offers me a little hedgehog of stamens, blue.

COLETTE, *Pour un Herbier*, 1948

Isn't Paradise an appendix to botany?

ÉMILE MICHEL CIORAN, *Le Crépuscule des Pensées*, 1940

TO END our comparison of plants and animals I would like to ask a very unscientific question: Which of the two kingdoms represents the greatest biological success? This is not a normal question, because it is unusual to ask something to which everyone knows the answer, and embarrassing because the answer, requiring consensus, has little chance of being objective.

For criteria of biological success, we will obviously choose those that place the animal in the position of conqueror—such as active movement, or sonorous utterance, or the famous criterion of reproductive success (fitness), the value of which has been investigated (Chapter 6)—providing the answer even before debate begins. At any rate, to choose criteria without having investigated their merits vis-à-vis the two kingdoms is contrary to fairness and rigor. Am I correct in suspecting a human bias about what is considered inferior?

What Do We Recognize as Success?

A study by Malcolm B. Wilkins (1995), a physiologist at the University of Glasgow, sheds light on the question of what constitutes success. He asked, "Are plants intelligent?" The reader of his work is prepared for an answer of yes by his solid arguments. Plants, he wrote, are capable of perceiving not only light but also the color of that light. They also detect gravity and know how to react to mechanical contact through electrical impulses similar to those that circulate in the nervous system of an animal. They know how to count, at least to two. The carnivorous Venus flytrap (*Dionaea muscipula*) closes its leafy trap after two stimulations of its sensory hairs. One stimulus is not enough but is remembered for a minute, pending the second one. The protein of the prey is digested by the flytrap. If the food is primarily sugar, its chemistry does not induce secretion of enzymes by the plant, which reopens its trap after 2–3 days. Plants have memory, albeit limited, a sense of taste, are capable of making noises, measuring temperature, rearranging themselves, distinguishing between enemies and friends, even transferring genes from one to another using the appropriate vectors.

Given this, Wilkins asks if plants are capable of making intelligent use of the continuous flux of information coming from the surrounding environment, just as we are. His answer, no, surprises me a bit. In spite of a sensory system at least as wide-ranging as ours, no, he says, plants are not intelligent. Why? Because their responses to external stimuli are not made by choice—their responses (though always correct) are programmed.

Is it a case, then, that the possibility of erring determines the value of a response, that humans are intelligent because they are capable of making mistakes? Is it well advised to recognize intelligence by mistakes rather than success? Is it not a good alternative in the warfare that is life to entrust one's destiny to an infallible program?

In the end, I agree with Wilkins that the notion of intelligence is not applicable to plants, that intelligence is specifically animal in nature. Intuitively, his verdict seems correct to me in that sense. Besides, what a disgrace for the herbivore to be less intelligent than the food in its stomach! But there is bias in a judgment rendered unilaterally by only one of the protagonists and based on a criterion tailored to assure victory, with its adversary incapable of speaking for itself. The plant is mute and refuses to hire a lawyer. Besides, the argument is of no interest to it. Still, a lawyer would have an easy task, and I would like to summarize how the case could be pleaded.

In Praise of Plants

That plants are different from animals is a commonplace. What is less recognized is that biology, founded on what we know of animals, takes practically no account of plants. This is partly the result of our lack of understanding of the plant world, in which big surprises remain to be discovered. It is also partly the result of biologists hesitating to accept novelty stemming from plants, as evidenced by the examples of genetic heterogeneity within a single tree or asexual transfer of genetic information between two different plant species. In addition to these examples, another has been added more recently. The well-known influence of the Moon on plants, recognized in popular culture, has long been denied by official science. But the influence of tides on trees has been demonstrated (Zürcher et al. 1998).

The result is a biology tailored to animals, including humans, while plants must submit to lying in a Procrustean bed. Ideas as fundamental as those of the individual, the genome, sexuality, the species, or evolution require modification or transformation to arrive at an objective biology, rid of zoocentrism and anthropocentrism. The situation is so much more paradoxical because there is evidence at every level establishing the superiority of plants over animals.

The plant cell is probably the most perfect of the two. It achieves a quasi-totality of functions of the animal cell to which it adds the key to all biology: photosynthesis. It also succeeds in keeping its totipotency, of which the animal cell is incapable. Plants are fixed in place, that is a fact, yet it signifies that they confront adversity instead of fleeing from it, as done so often by animals. Consequently, plants have had to develop enormous capacities for resistance, of which a large part comes from their genetic plasticity. A poorly integrated organism, the plant benefits from the fact that it is, according to Tsvi Sachs of the University of Jerusalem, a population of redundant organs competing with one another (Sachs et al. 1993), promoting that part of the genome better adapted to the conditions of the moment. If conditions change, plants put a variant of the initial genome to work, one better adapted to the new environment.

Dispersion or Concentration

From all the evidence, animals have attained a high level of sophistication in structure, function, and behavior, but make no mistake, it is at the price of continuous dispersion of energy obtained by preying on plants. It is not difficult to live a dazzling life by squandering the resources of others, and as we know, *la dolce vita* is not the lot of the hardworking. Plants constitute 99% of the biomass of multicellular organisms, working all the time and with great sobriety. They achieve this with only three types of organs (root, stem, leaf), using some water and a bit of waste for a skeleton—and this building process is effected through the concentration of energy (Plate 99).

Moreover, plants are not simple structurally. They, too, attain a high level of sophistication in perception of external stimuli and in cooperation between organs. It is true that they are silent; they communicate in other ways. Plants are the largest living beings and those that live the longest. They are also at the origin of most food chains. They constitute the structure of our landscapes, shelter animals, help form soils, control local climate, refresh the atmosphere, and are even capable of cleaning up pollutants to a certain extent. Thanks to photosynthesis, plants furnish animals with energy, food, and air.

The biological success of plants is demonstrated nowhere better than in feeding: Plants do not depend on animals, whereas animals depend on plants for their daily survival. Humans, too, depend completely on plants for food. It matters little whether we are vegetarian

Figure 99. Dispersion or concentration. Plants concentrate energy; animals disperse it.

or not. Without plants, I believe it would be necessary to nourish ourselves on water and salt! Humans, the self-proclaimed summit of evolution . . .

In their relationships with animals, plants are not content to play the passive role of food without defending themselves. When they need to, plants know how to borrow the mobility of animals, their quickened temporal rhythms, their globular forms, and their disgusting odors. As well-informed manipulators, they know how to exploit the weaknesses of their partners, achieving their own ends through collaboration.

To my knowledge, these are uncontradicted facts. Still, this recitation is not completely realistic because even our language is not appropriate for plants. They do not *know*. They *use* nothing. They have no *needs, projects,* or *goals.* We speak an animal language that adapts poorly to the truth about plants. These anthropomorphisms are preferable to tedious paraphrasing, however, and we hardly have a choice. The language of plants, if they had one, would perhaps be a little slow. At least our language allows us celebrate the aesthetic qualities of plants. Plants are beautiful. They smell good, dead or alive. We owe a portion of our mental equilibrium to them, just as we note their importance in our cities. A large part of the world's beauty is because of them. In my opinion, this is more than simple beauty.

Are Plants Persons?

A stupid question, certainly, and it is giving in to obscurantism to ask it. Above all, it is an imprudent and difficult question, one that I make no pretense of answering. I only wish to sidelight it a little, to provoke some reflection.

In the humid Tropics, members of forest tribes are unanimous in considering plants as, in a certain sense, persons. These beliefs, propagated by shamans, have been reported among the Embera of the Chocó in Colombia (Gómez Díaz 1996) and the Shuar in Ecuador (Descola 1996). Such beliefs are generally held by the Amerindians of low-elevation forests in South America (Darrel Posey, personal communication). In Europe, such an idea is shocking. But who must believe in it: Westerners who deny the personalities of plants without ever having given them much attention, or shamans who have lived their entire lives in contact with the most diverse floras on the planet? The latter have formed close connections with thousands of plants, who for them are like family and who have become their accomplices. I am not aware if plants have true personality, but I ask myself, Who has the most authority on such a subtle topic?

I do not have much to add to this debate—only this, what those who cultivate plants for pleasure know well: Although plants do not defend themselves when they are harmed, they are sensitive to the attention, sometimes clumsy, that they receive. Add a little common sense and respect, some good soil, a bit of warmth, water, and light, in exchange for a magical result. Happily, many of our contemporaries are grateful for such gifts and agree with these thoughts. Like Émile Cioran, have you noticed that a botanical garden is often cast in the image of a paradise? A zoo, whatever one makes of it, rather resembles a hell.

The Two Faces of Botany

Botany with a capital B is collection, conservation, and analysis in museums. It is those who publish Latin diagnoses and DNA sequences, curves and graphs, in specialized journals using only the most rigorous if inelegant terminology. It is those who present their results at congresses in front of an auditorium filled with colleagues who hardly listen, waiting for their turn to speak. Botany does not enjoy the image of enlightenment, evoking instead the image of the ivory tower, esoteric and dusty, rather than the garden of Eden. Not unexpectedly, this official Botany is shrinking, the administrative powers not replacing its teachers when they disappear, no longer paying for the care of its historical collections, allowing instead the invasion of its laboratories by more aggressive and marketable enterprises. It is characteristic of our time that this Botany is unloved by the technocrats who manage scientific research; they prefer more lucrative and shorter-term activities. Its future is bleak, marked by the melancholy of its worn-out practitioners. In France, this official Botany is in the process of disappearing while the classrooms are full of students with a passion for plants.

Very happily, there is another botany, unofficial and alternative, one fully alive and abundant, burgeoning, suckering, polymorphous, and multicolored, eclectic, unclassifiable, erotic, pagan, and inconceivable to research managers, institutional heads, and bureaucrats. This botany is not very orthodox in the eyes of official science, not very concerned with clearly drawn boundaries, and is in fact poorly distinguished from agricultural biology, militant environmentalism, homeopathy, and the art of living, gastronomy, and magic. This botany quite simply belongs to those who love plants: to Pierre Lieutaghi and his medieval gardeners, to Gilles Clément and his moving garden, to René Hebding and his marvelous greenhouses, to Patrick Blanc and his gardens on walls, to Maurice Chaudière and his fruit-bearing garrigue. It includes bonsai and ikebana, journeys in *le radeau des cimes* (the canopy raft), *Garance Voyageuse*, an association and journal of botanical popularization, and *La Hulotte*, a journal devoted to animals and plants. The human species needs plants, especially in these times of degraded environments. This is why, in one form or another, official or alternative, knowledge of plants is resurgent in our culture despite the commotion it causes, as a dandelion keeps reappearing in the same place despite attempts to remove it.

It is the entire plant, from its roots to its flowers, in its soil, with its uses through the ages, that is important, because we need to perceive it with our senses, not only in an intellectual, devitalized way. The specialists often do not understand this, reducing the plant to a chromosome number, a sequence of base pairs, a Latin binomial, an electronic image of an organelle, a point on a curve, a bibliographic reference, a datum in a computer memory, a centrifuged residue, or a callus at the bottom of a test tube.

What is going on? The crowd observes, with the respect due to official science, but is distracted. More interesting is the truth that attracts the gardener and his or her accomplices, including the horticulturist and the herbalist, the grape grower and the healer, the landscape painter and the poet. They do not doubt their responsibility for maintaining a very ancient knowledge of life, full of the future and indissolubly linked to the human species.

I am not satisfied with a two-faced botany. I would like to see a unified science, open to the entire world, spurning obscurantism, rigorous but without jargon, accepting of amateur abilities, capable of satisfying the passions of the very young, restoring the primacy of observation, associating fieldwork with research in the laboratory, combining plant geography with the study of fossils, ethnobotany with mathematics, genetics with the analysis of form, knowing how to call on the most advanced biological techniques without repudiating the value of traditional tools. I dream of a botany that determines its own future, according to its own rules. No longer the caboose at the end of the train of animal and human physiology, considering the plant itself as an archetypal life form, as a model for its own study and that of the environment, botany could reestablish its place at the center of the life sciences.

When degradation in the quality of life disturbs us, when the trio concrete-asphalt-automobile takes on the color of hell and the smell of the sewer, we should be inspired by plants, by their sobriety, their prudence, their dignity—the future for our planet would seem less gloomy. In our world of pretence, money, commercials, noise, pollution, and violence, what better testimony than that of plants, beautiful and useful, reserved and self-abiding, silent and completely nonviolent?

References

Alberts, B., D. Bray, et al. 1983. The Molecular Biology of the Cell. Garland, New York.

Andrews, J. H. 1991. Comparative Ecology of Microorganisms and Macroorganisms. Springer-Verlag, Heidelberg.

Antolin, M. F., and C. Strobeck. 1985. The population genetics of somatic mutation in plants. American Naturalist 126: 52–62.

Appia, D. 1977. Le Grand Voyage [poster]. Wizard & Genius, Meilen, Switzerland.

ASSIM (Association des Enseignants d'Immunologie des Universités de Langue Française). 1991. Immunologie Générale, éd. 2. MEDSI–McGraw-Hill, Paris.

Atger, C., and C. Edelin. 1994. Strategies d'occupation du milieu souterrain par les systèmes racinaires des arbres. La Terre et la Vie 49: 343–356.

Baldry, J., J. Dougan, and G. E. Howard. 1972. Volatile flavoring constituents of durian. Phytochemistry 11: 2081–2084.

Barlow, B. A., and D. Wiens. 1977. Host–parasite resemblance in Australian mistletoes: the case for cryptic mimicry. Evolution 31: 69–84.

Barnes, J. (editor). 1984. The Complete Works of Aristotle, 2 vols. Princeton University Press, Princeton.

Barnsley, M. 1985. Iterated function systems and the global construction of fractals. Proceedings of the Royal Society of London, A, 399: 243–275.

Barthélémy, D. 1983. Convergences Géographiques. L'Hypothèse d'une Transmission d'Information Génétique sur des Voies Non Sexuelles. Thèse de D.E.A., Université Montpellier II.

Bates, H. W. 1862. Contributions to an insect fauna of the Amazon Valley, Lepidoptera: Heliconidae. Transactions of the Linnean Society of London 23: 495–515.

Baudelaire, C. 1998. Selected Poems from *Les Fleurs du Mal*, translated by N. R. Shapiro. University of Chicago Press, Chicago.

Beklemishev, V. N. 1970. Principles of the Comparative Anatomy of Invertebrates. University of Chicago Press, Chicago.

Bennett, M. D., J. B. Smith, and J. S. Heslop-Harrison. 1982. Nuclear DNA amounts in angiosperms. Proceedings of the Royal Society of London, B, 216: 179–199.

Bernard, C. 1878–79. Leçons sur les Phénomènes de la Vie Communs aux Animaux et aux Végétaux, 2 vols. J. B. Baillière, Paris.

Bernard, J., M. Bessis, and C. Debru (editors). 1990. Soi et Non-soi. Éditions du Seuil, Paris.

Bertalanffy, L. von. 1960. Principles and theory of growth, pages 137–259 in W. W. Nowinski (editor), Fundamental Aspects of Normal and Malignant Growth. Elsevier, Amsterdam.

Bertin, L. 1950. La Vie des Animaux, 2 vols. Larousse, Paris.

Biaggi, V., and J. Arnaud. 1995. Poulpes, Seiches et Calmars: Mythes et Gastronomie. Jeanne Laffitte, Paris.

Bierne, J. 1994. De la biologie de la régéneration à la biotechnologie de la embryogenèse somatique chez les animaux. Communication, Embryotech '94, Lyon, 25–26 October.

Blunt, W. 1971. The Compleat Naturalist, a Life of Linnaeus. Collins, London.

Boardman, R. S., A. H. Cheetham, and W. A. Oliver (editors). 1973. Animal Colonies, Development and Function Through Time. Dowden, Hutchinson and Ross, East Stroudsburg, Pennsylvania.

Bonner, J. T., A. D. Chiquoine, and M. Q. Kolderie. 1955. A histochemical study of differentiation in the cellular slime molds. Journal of Experimental Zoology 130: 133–158.

Borges, J. L., and M. Guerrero. 1957. Manual de Zoología Fantástica. Mexico. (1969. The Book of Imaginary Beings, revised, enlarged, and translated by N. T. di Giovanni in collaboration with the author. Dutton, New York.)

Bourdu, R., and M. Viard. 1988. Arbres Souverains. Éditions du May, Paris.

Boutot, A. 1993. L'Invention des Formes. Odile Jacob, Paris.

Bownas, G. 1964. The Penguin Book of Japanese Verse. Penguin Books, Baltimore.

Bradshaw, A. D. 1965. Evolutionary significance of phenotypic plasticity in plants. Advances in Genetics 13: 115–155.

Bradshaw, A. D. 1972. Some of the evolutionary consequences of being a plant. Evolutionary Biology 6: 25–47.

Brosse, J. 1958. L'Ordre des Choses. Plon, Paris.

Brosse, J. 1989. Mythologie des Arbres. Plon, Paris.

Buffon, G. L. 1971. De l'Homme. Maspero, Paris.

Burnet, F. M. 1969. Cellular Immunology. Self and Non-self. Cambridge University Press, Cambridge.

Burrows, C. J., B. McCulloch, and M. M. Trotter. 1981. The diet of moas. Records of the Canterbury Museum 9: 309–336.

Buss, L. W. 1983. Evolution, development, and the units of selection. Proceedings of the National Academy of Sciences U.S.A. 80: 1387–1391.

Buss, L. W. 1987. The Evolution of Individuality. Princeton University Press, Princeton.

Cairns-Smith, A. G. 1971. The Life Puzzle: on Crystals and Organisms and on the Possibility of a Crystal as an Ancestor. Oliver and Boyd, London.

Calvino, I. 1959. The Baron in the Trees. Random House, New York.

Cather, W. 1913. O Pioneers! Houghton Mifflin, Boston.

Cendrars, B. 1924. Au Coeur du Monde. Sagittaire, Paris.

Champagnat, P. 1987. Deux aspects du développement des végétaux, pages 3–21 in H. Le Guyader (editor), Développement des Végétaux: Aspects Théoriques et Synthétiques. Masson, Paris.

Charlesworth, D. 1989. A high mutation rate in a long lived perennial plant. Nature 340: 346–347.

Chaudière, M. 1992. Créons des forêts fruitières. Les Quatres Saisons du Jardinage 77: 19–22.

Chivers, D. J., and C. M. Hladik. 1980. Morphology of the gastrointestinal tract in primates: comparisons with other mammals in relation to diet. Journal of Morphology 166: 337–386.

Cioran, É. M. 1973. De l'Inconvénient d'Être Né. Gallimard, Paris.

Cioran, É. M. 1995. Le Crépuscule du Pensées. Gallimard, Paris.

Cline, M. G. 1997. Concepts and terminology of apical dominance. American Journal of Botany 84: 1064–1069.

Coates, A. G., and J. B. C. Jackson. 1985. Morphological themes in the evolution of clonal and aclonal marine invertebrates, pages 67–106 in J. B. C. Jackson, L. W. Buss, and R. E. Cook (editors), Population Biology and Evolution of Clonal Organisms. Yale University Press, New Haven.

Coen, E. S., and E. M. Meyerowitz. 1991. The war of the whorls: genetic interactions controlling flower development. Nature 353: 31–37.

Colette, S.-G. 1948. Pour un Herbier. Mermod, Lausanne.

Coll, J. C., S. La Barre, P. Sammarco, W. T. Williams, and G. J. Backus. 1982. Chemical defenses in soft corals of the Great Barrier Reef: a study of comparative toxicities. Marine Ecology Progress Series 8: 271–278.

Combes, C. 1995. Interactions Durables: Écologie et Évolution du Parasitism. Masson, Paris. (2001. Parasitism: the Ecology and Evolution of Intimate Interactions, translated by I. de Buron and V. A. Connors. University of Chicago Press, Chicago.)

Comfort, A. 1956. The Biology of Senescence. Routledge & Paul, London (ed. 3, 1979, Elsevier, New York).

Cook, C. D. K. 1974. Water Plants of the World. W. Junk, The Hague.

Corbin, A. 1986. Le Miasme et la Jonquille. Flammarion, Paris.

Corliss, J. B., et al. 1979. Submarine thermal springs on the Galápagos Rift. Science 203: 1073–1075.

Corner, E. J. H. 1949. The durian theory, or the origin of the modern tree. Annals of Botany, new ser., 13 (52): 367–414.

Corner, E. J. H. 1958. Transference of function. Botanical Journal of the Linnean Society of London 56: 33–40.

Couvens, R., and M. J. Hutchings. 1983. The relationship between density and mean frond weight in a monospecific sea weed stand. Nature 301: 240–241.

Crow, J. F., and M. Kimura. 1970. An Introduction to Population Genetics Theory. Harper & Row, New York.

Cruiziat, P., and H. Le Guyader. 1990. Un exemple de fonction en physiologie végétale: les relations plante–eau. École Européenne de Biologie Théoretique, Solignac, 2–22 Septembre.

Cullis, C. A. 1986. Phenotypic consequences of environmentally induced changes in plant DNA. Trends in Genetics 2(12): 307–309.

D'Arcy Thompson, W. 1917. On Growth and Form. Cambridge University Press, Cambridge.

Darley, M. W. 1990. The essence of "plantness." American Biology Teacher 52: 354–357.

Darmency, H. 1994. Genetics of herbicide resistance in weeds and crops, pages 263–297 in S. B. Powles and J. A. M. Holttum (editors), Herbicide Resistance in Weeds and Crops: Biology and Biochemistry. Lewis, Boca Raton.

Dauchez, A. 1900. L'Histoire de l'Homme Goémon. Unpublished, of the Hallé family.

Dauget, J.-M. 1985. La réaction aux traumatismes: comparaison entre les arbres et les coraux. La Terre et La Vie 40: 113–118.

Dauget, J.-M. 1991. Application of tree architectural models to reef-coral growth forms. Marine Biology 111: 157–175.

Dausset, J. 1990. La définition biologique du soi, pages 19–26 in J. Bernard, M. Bessis, and C. Debru (editors), Soi et Non-soi. Éditions du Seuil, Paris.

Dawkins, R. 1995. God's utility function. Scientific American 273(5): 80–85.

Dawson, J. W. 1963. A comment on divaricating shrubs. Tuatara 11: 193–194.

Dekeyser, W., and S. Amelinckx. 1955. Les Dislocations et la Croissance des Cristaux. Masson, Paris.

Descola, P. 1996. Les cosmologies des Indiens d'Amazonie. La Recherche 292: 62.

Dittmer, H. J. 1937. A quantitative survey of the roots and root hairs of a winter rye plant (Secale cereale). American Journal of Botany 24: 417–420.

Doffin, H. 1959. La phyllotaxie des caractères crystallographiques des végétaux en tant que corps solides réguliers. Travaux du Laboratoire de Biologie Végétal de Faculté Sciences Poitiers 14: 3–14.

Douady, S., and Y. Couder. 1992. Phyllotaxis as a physical self-organized growth process. Physical Review Letters 68: 2098–2101.

Douady, S., and Y. Couder. 1993. Les spirales vegetales. La Recherche 24 (250): 26–35.

Douglas, S. 1998. Plastid evolution: origins, diversity, trends. Current Opinions in Genetics and Development 8: 655–661.

Doumenc, D., and P. M. Lenicque. 1995. La Morphogenèse: Développement et Diversité des Formes Vivantes. Masson, Paris.

Downum, K. R., D. W. Lee, F. Hallé, M. Quirke, and N. Towers. 2001. Plant secondary compounds in the canopy and understorey of a tropical rain forest in Gabon. Journal of Tropical Ecology 17: 477–481.

Ducreux, G., and H. Le Guyader. 1995. L'invention de la cellule méristématique. Bulletin de la Société Zoologique Française 120: 139–155.

Duhamel du Monceau, H. L. 1758. La Physique des Arbres. H. L. Guerin et L. F. Delatour, Paris.

Dumortier, B. C. J. 1832. Recherches sur la Structure Comparée et la Développement des Animaux et des Végétaux. H. Hayez, Brussells.

Du Trochet, H. 1824. Recherches Anatomiques et Physiologiques sur la Structure Intime des Animaux et des Végétaux, et sur Leur Motilité. J. B. Baillière, Paris.

Du Trochet, H. 1826. L'Agent Immédiate du Mouvement Vital Devoilé des Sa Nature et dans Son Mode d'Action Chez les Végétaux et Chez les Animaux. J. B. Baillière, Paris.

Dyrnda, P. E. J. 1986. Defensive strategies of modular organisms. Philosophical Transactions of the Royal Society of London, B, 313: 227–243.

Dyson, F. 1971. Energy in the universe. Scientific American 225(9): 50–59.

Edelin, C. 1990. Monopodial architecture: the case of some tree species from Southeast Asia. Forest Research Institute of Malaysia, Kuala Lumpur, Research Pamphlet 105.

Ehrlich, G. 1988. River history, pages 69–72 in William Kittredge (editor), Montana Spaces: Essays and Photographs in Celebration of Montana. Nick Lyons Books, New York.

Eliade, M. 1949. Le Mythe de l'Éternel Retour. Gallimard, Paris. (1954. The Myth of the Eternal Return, or, Cosmos and History, translated by W. R. Trask. Princeton University Press, Princeton.)

Errera, L. 1910. Exposés de Physiologie Générale.

Fabre, J.-H. 1996. La Plante, Leçons à Mon Fils sur la Botanique. Privat, Paris.

Favre, C., and D. Favre. 1991. Naissance du Quatrième Type. Barret-le-Bas.

Favre-Duchartre, M. 1997. Unions Créatrices: Point de Vue d'un Naturaliste Teilhardien. Éditions Aubin, Saint Etienne.

Fincham, J. R. S. 1983. Genetics at first sight. Nature 304: 377–378.

Fitzroy, R. 1839. Narrative of the Surveying Voyages of His Majesty's Ships *Adventure* and *Beagle*, Between the Years 1826 and

1836, Describing Their Examination of the Southern Shores of South America, and the *Beagle*'s Circumnavigation of the Globe. H. Colburn, London.

Franc, A. 1968. Traité de Zoologie, vol. 5, 3 Mollusques Opisto-branches. Masson, Paris.

Friedmann, F., and T. Cadet. 1976. Observations sur l'heterophyllie dans les îles Mascareignes. Adansonia, sér. 2, 15: 423–440.

Funk, V. A. 1981. Special concerns in estimating plant phylogenies, pages 73–86 in V. A. Funk and D. R. Brooks (editors), Advances in Cladistics: Proceedings of the First Meeting of the Willi Hennig Society. New York Botanical Garden, New York.

Gidoin, J.-M. 1991. Contes pour Nos Enfants. Sélection du Reader's Digest, Paris.

Gilbert, L. E. 1975. Ecological consequences of a coevolved mutualism between butterflies and plants, pages 210–240 in L. E. Gilbert and P. H. Raven (editors), Coevolution of Plants and Animals. University of Texas Press, Austin.

Gilbert, L. E. 1991. Evolutionary responses of *Passiflora* to *Heliconius* attack, pages 402–427 in W. P. Price, T. M. Lewinson, G. W. Fernandez, and W. W. Benson (editors), Plant–Animal Interactions. John Wiley, New York.

Gill, D. E. 1986. Individual plants as genetic mosaics: ecologial organisms versus evolutionary individuals, pages 321–343 in M. J. Crawley (editor), Plant Ecology. Blackwell, Oxford.

Gill, D. E., and T. G. Halverston. 1984. Fitness variation among branches within trees, pages 105–116 in B. Shorrocks (editor), Evolutionary Ecology. Blackwell, London.

Goethe, W. 1974. The Autobiography of Johann Wolfgang von Goethe, vol. 1, translated by J. Oxenford. University of Chicago Press, Chicago.

Gómez Díaz, J. A. 1996. Ethnobotanique de Trois Communautés Amérindiennes Embera dans la Région Pacifique de Colombie. Thèse, Université Monpellier II.

Goreau, T. F., N. Goreau, and T. J. Goreau. 1979. Coraux et récifs coralliens. Pour la Science 24: 77–88.

Gottlieb, L. D. 1984. Genetics and morphological evolution in plants. American Naturalist 123: 681–709.

Gould, S. J. 1987. An Urchin in the Storm: Essays About Books and Ideas. W. W. Norton, New York.

Gould, S. J. 1989. Wonderful Life: the Burgess Shale and the Nature of History. W. W. Norton, New York.

Gould, S. J. (editor). 1993. The Book of Life. W. W. Norton, New York.

Gould, S. J. 1996. Full House: the Spread of Excellence from Plato to Darwin. Harmony Books, New York.

Gouyon, P.-H. 1996. Le Néodarwinisme ne menace pas l'ethique: les individus sont des artifices inventés par les gènes pour se reproduire. La Recherche 292(11): 88–92.

Grassé, P.-P. (editor). 1948–96. Traité de Zoologie: Anatomie, Systématique, Biologie. Masson, Paris.

Grassé, P.-P. 1973. L'Évolution du Vivant: Matériaux pour une Nouvelle Théorie Transformiste. Albin Michel, Paris.

Green, G. 1977. Ecology of toxicity in marine sponges. Marine Biology 40: 207–215.

Greene, E. 1989. A diet-induced developmental polymorphism in a caterpillar. Science 243: 643–646.

Greenwood, R. M., and I. A. E. Atkinson. 1977. Evolution of divaricating plants in New Zealand, in relation to moa browsing. Proceedings of the New Zealand Ecological Society 24: 21–33.

Grime, J. P., J. C. Crick, and J. E. Rincon. 1986. The ecological significance of plasticity, pages 5–25 in D. H. Jennings and A. J. Trewavas (editors), Plasticity in Plants. Cambridge University Press, Cambridge.

Gudin, C. 1996. Nique Ta Botanique. L'Âge d'Homme, Lausanne, Switzerland.

Haber, A. H., W. L. Carrier, and D. E. Ford. 1961. Metabolic studies of gamma-irradiated wheat growing without cell divisions. American Journal of Botany 48: 431–438.

Haeckel, E. 1897. The Evolution of Man, 2 vols. D. Appleton and Company, New York.

Hagemann, W. 1982. Vergleichende Morphologie und Anatomie. Organismus und Zelle: Ist eine Synthese Mögliche? Berichte der Deutschen Botanischen Gesellschaft 95: 435–446.

Hales, S. 1727. Vegetable Staticks. London. (Facsimile: 1969. American Elsevier, New York.)

Hallé, F. 1993. Un Monde sans Hiver: les Tropiques, Nature et Sociétés. Éditions du Seuil, Paris.

Hallé, F., and R. A. A. Oldeman. 1970. Essai sur l'Architecture et la

Dynamique de Croissance des Arbres Tropicaux. Masson, Paris. (1975. An Essay on the Architecture and Dynamics of Growth of Tropical Trees, translated by B. C. Stone. Kuala Lumpur, Penerbit Universiti Malaya.)

Hallé, F., R. A. A. Oldeman, and P. B. Tomlinson. 1978. Tropical Trees and Forests: an Architectural Analysis. Springer-Verlag, Berlin.

Harper, J. L. 1977. Population Biology of Plants. Academic Press, London.

Harper, J., B. R. Rosen, and J. White. 1986. The growth and form of modular organisms. Philosophical Transactions of the Royal Society of London, B, 313: 3–5.

Hashimoto, Y. 1979. Marine Toxins and Other Bioactive Marine Metabolites. Japan Scientific Societies Press, Tokyo.

Hawkins, C. 1990. Les Monstres. Albin Michel, Paris.

Heller, J. L. 1964. The early history of biological nomenclature. Huntia 1: 33–70.

Hendrix, S. D. 1988. Herbivory and its impact on plant reproduction, pages 246–263 in J. and L. Lovett-Doust (editors), Plant Reproductive Ecology: Patterns and Strategies. Oxford University Press, New York.

Heywood, V. H. (editor). 1985. Flowering Plants of the World. Croom Helm, London.

Hladik, C. M. 1967. Surface relative du tractus digestif de quelques primates: morphology des villosités intestinales et corrélations avec la régime alimentaire. Mammalia 31: 120–147.

Holldobler, B., and E. O. Wilson. 1994. Journey to the Ants: a Story of Scientific Exploration. Harvard University Press, Cambridge.

Hugo, V. 1883 La Légendes des Siècles (1979: Flammarion, Paris).

Humphreys, T. 1970. Biochemical analysis of sponge cell aggregation. Zoological Society of London Symposium 25: 325–334.

Huygen, W. 1979. Les Gnomes. Albin Michel, Paris.

Hylands, P. J. 1994. Maximizing diversity in plant based drug discovery, in Drugs from Nature, a Conference, London, 8–9 December.

Ingber, D. E. 1998. The architecture of life. Scientific American 278(1): 48–57.

Jackson, J. B. C., L. W. Buss, and R. E. Cook. 1985. Population Biology and Evolution of Clonal Organisms. Yale University Press, New Haven.

Jacob, F. 1970. La Logique du Vivant: une Histoire de l'Hérédité. Gallimard, Paris. (1973. The Logic of Life: a History of Heredity, translated by B. E. Spillmann. Pantheon, New York.)

Jaeger, J.-J. 1996. Les Mondes Fossiles. Odile Jacob, Paris.

Jaubert, J.-N. 1987. Chimie, parfums et arômes végétaux, in Parfums des Plantes. Museum National d'Histoire Naturelle, Paris.

Jennings, D. H., and A. J. Trewavas (editors). 1986. Plasticity in Plants: Symposium for Experimental Biology. Cambridge University Press, Cambridge.

Jorodowsky, K., and Moebius. 1981. L'Incal Noir: une Aventure de John Difool. Les Humanoïdes Associés, Paris.

Jourdan, C. 1995. Modélisation de l'Architecture et du Développement du Palmier à Huile. Thèse, Université Montpellier II.

Jullien, M. 1980. Les cultures cellules chez les végétaux supériors et leurs applications, pages 161–190 in R. Chaussat and C. Bigot (editors), La Multiplication Végétative des Plantes Supérieures. Gauthier-Villars, Paris.

Jurgens, G. 1992. Genes to greens: embryonic pattern formation in plants. Science 256: 487–488.

Kaplan, D. R. 1992. The relationship of cells to organisms in plants: problem and implications of an organismal perspective. International Journal of Plant Sciences 153: S28–S37.

Kaplan, D. R., and W. Hagemann. 1991. The relationship of cell and organism in vascular plants: Are cells the building blocks of plant form? BioScience 41: 693–703.

Keller, E. F. 1983. A Feeling for the Organism: the Life and Work of Barbara McClintock. W. H. Freeman, New York.

Kellert, S. R., and E. O. Wilson (editors). 1993. The Biophilia Hypothesis. Island Press, Washington, D.C.

Kenyon, J. C. 1992. Chromosome numbers in ten species of the coral genus *Acropora*. Proceedings of the 7th International Coral Reef Symposium 1: 471–475.

Kessler, B., and S. Reches. 1977. Structural and functional changes of chromosomal DNA during aging and phase change in plants. Chromosomes Today 6: 237–246.

King, J. 1997. Reaching for the Sun: How Plants Work. Cambridge University Press, Cambridge. [See also Pollan (2001), a book describing how plants used secondary metabolites to manipulate human desire for sweetness, beauty, intoxication, etc., to increase their own abundance, and fitness. If the author had seen that book, which was published after his, he would have quoted it. — translator's comment]

Kittel, C. 1958. Introduction à la Physique de l'État Solide. Dunod, Paris.

Klekowski, E. J. 1988. Mutation, Developmental Selection, and Plant Evolution. Columbia University Press, New York.

Klekowski, E. J., and N. Kazarinova-Fukshansky. 1984a. Shoot apical meristems and mutation: fixation of selectively neutral cell genotypes. American Journal of Botany 71: 22–27.

Klekowski, E. J., and N. Kazarinova-Fukshansky. 1984b. Shoot apical meristems and mutation: selective loss of disadvantageous cell genotypes. American Journal of Botany 71: 28–34.

Klekowski, E. J., N. Kazarinova-Fukshansky, and L. Fukshansky. 1989. Patterns of plant ontogeny that may influence genomic stasis. American Journal of Botany 76: 189–195.

Knight-Jones, E. W., and J. Moyse. 1961. Intraspecific competition in sedentary marine animals. Society of Experimental Biology Symposium 15: 72–95.

Kondrashov, A. S., and J. F. Crow. 1991. Haploidy or diploidy: Which is better? Nature 351: 314–315.

Kullenberg, B., and G. Bergstrom. 1976. Hymenoptera aculeata males as pollinators of Ophrys orchids. Zoological Scripta 5: 13–23.

Laborel, J. 1970. Madréporaires et hydrocoralliaires récifaux des côtes brésiliennes. Résultats Scientifiques des Campagnes de la Calypso. Masson, Paris.

Lacarrière, J. 1985. Lapidaire: Suivi de, Lichens. Éditions Fata Morgana, Saint-Clément-de-Rivière.

La Fontaine, J. de. 1668–94. Fables. (1988. The Complete Fables of Jean de La Fontaine, translated by N. B. Spector. Northwestern University Press, Evanston, Illinois.)

Landman, O. E. 1991. The inheritance of acquired characteristics. Annual Reviews of Genetics 25: 1–20.

Laplace, Y. 1977. Cinq semaines auprès des "faux" de Verzy. Rapport de Stage, École Nationale du Génie Rural et des Eaux et Forêts, Nancy.

Larpent, J.-P. 1970. De la Cellule à l'Organisme: Acrasiales, Myxomycètes, Myxobactériales. Masson, Paris.

Larson, D. W., P. E. Kelly, and U. Matthes-Sears. 1995. Calling nature's bluff: the secret world of cliffs, pages 26–47 in Yearbook of Science and the Future. Encyclopaedia Brittanica, Chicago.

Lear, E. 1888. Nonsense Botany, and Nonsense Alphabets, etc., etc. F. Warne, London.

Leavitt, R. G. 1909. A vegetative mutant, and the principle of homeosis in plants. Botanical Gazette 47: 30–58.

Le Guin, U. K. 1972. The word for world is forest, in Again, Dangerous Visions. Berkley Publishing, New York.

Le Guyader, H. 1994. Évolution des complexes homéotiques et récapitulation. Bulletin de la Societé Zoologie Française 119: 127–148.

Leistikow, K. U. 1994. Plant construction related to plant reproduction, pages 177–180 in Evolution of Natural Structures: Principles, Strategies, and Models in Architecture and Nature. Proceedings of the III. International Symposium of the Sonderforschungsbereich 230. Universität Stuttgart und Universität Tübingen.

Leopold, A. C. 1978. The biological significance of death in plants, pages 101–114 in J. A. Behnke (editor), The Biology of Aging. Plenum, New York.

Lewis, W. H. 1970a. Chromosomal drift, a new phenomenon in plants. Science 168: 115–116.

Lewis, W. H. 1970b. Extreme instability of chromosome number in *Claytonia virginica*. Taxon 19: 180–182.

Libbert, E., and R. Manteuffel. 1970. Interactions between plants and epiphytic bacteria regarding their auxin metabolism. Physiologia Plantarum 23: 93–98.

Lieutaghi, P. 1991. La Plante Compagne: Pratique et Imaginaire de la Flore Sauvage en Europe Occidentale. Jardin Botanique de Genève, Genève.

Linnaeus, C. 1749. Specimen Academicum de Oeconomia Naturae. Uppsala.

Lintilhac, P. M. 1984. Positional controls in meristem development: a caveat and an alternative, pages 83–105 in P. J. Barlow and D. J. Carr (editors), Positional Controls in Plant Development. Cambridge University Press, New York.

Luneau, M. 1994. Paroles d'Arbre. Julliard, Paris.

Ma, H. 1994. The unfolding drama of flower development: recent results from genetic and molecular analyses. Genes & Development 8: 745–756.

McClintock, B. 1984. The significance of responses of the genome to challenge. Science 226: 792–801.

McLaren, N. 1947. La Poulette Grise [animated movie]. National Film Board of Canada, Ottawa.

Mandela, N. 1994. Long Walk to Freedom. Little, Brown and Company, Boston.

Margara, J. 1982. Bases de la Multiplication Végétative: les Méristèmes et l'Organogenèse. INRA, Paris.

Margulis, L., and R. Fester. 1991. Symbiosis as a Source of Evolutionary Innovation: Speciation and Morphogenesis. MIT Press, Cambridge.

Margulis, L., and K. V. Schwartz. 1982 (ed. 2, 1988). Five Kingdoms: an Illustrated Guide to the Phyla of Life on Earth. W. H. Freeman, New York.

Martin, J. P., and I. Fridovich. 1981. Evidence for a natural gene transfer from ponyfish to its bioluminescent bacterial symbiont *Photobacter leiognathi*. Journal of Biological Chemistry 256: 6080–6089.

Martínez, E., and B. Santelices. 1992. Size hierarchy and the $-\frac{3}{2}$ "power law" relation in the coalescent sea weed *Iridaea laminarioides*. Journal of Phycology 28: 259–264.

Masclef, A. 1891. Atlas des Plantes de France: Utiles, Nuisables et Ornementales. Paul Klincksieck, Paris.

Matile, P. 1975. The Lytic Compartment of Plant Cells. Springer-Verlag, Vienna.

Meins, F. 1983. Heritable variation in plant cell culture. Annual Reviews of Plant Physiology 34: 327–346.

Meyer, A. 1987. Phenotypic plasticity and heterochrony in *Cichlasoma managuense* (Pisces, Cichlidae) and their implications for speciation in cichlid fishes. Evolution 41: 1357–1359.

Meyerowitz, E. 1994. The genetics of flower development. Scientific American 271(11): 56–65.

Mezières, J.-C., and P. Christin. 1971. L'Empire des Mille Planètes. Dargaud, Paris.

Michaux-Ferrière, N. 1974. Évolution du fonctionnement méristématique au cours du développement chez *Pteris cretica* L. Apports des moyens d'étude modernes. Saussurea 5: 49–60.

Michel, F.-B. 1999. La Face Humaine de Vincent van Gogh. Éditions Grasset, Paris.

Moffett, M. 1995. The High Frontier: Exploring the Tropical Rainforest Canopy. Harvard University Press, Cambridge.

Monod, J. 1956. Remarks on the mechanism of enzyme induction, pages 7–28 in O. H. Gaebler (editor), Enzymes—Units of Biological Structure and Function. Henry Ford Hospital International Symposium, Detroit.

Monod, J. 1970. Le Hasard et la Nécessité: Essai sur la Philosophie Naturelle de la Biologie Moderne. Éditions du Seuil, Paris. (1971. Chance and Necessity: an Essay on the Natural Philosophy of Modern Biology, translated by A. Wainhouse. Knopf, New York.)

Murawski, D. A. 1998. Genetic variation within tropical tree crowns, pages 104–113 In F. Hallé (editor), Biologie d'une Canopée de Forêt Tropicale. Pro-Natura International et Operation Canopée, Paris.

Myers, N., R. A. Mittermeier, C. G. Mittermeier, G. A. B. da Fonseca, and J. Kent. 2000. Biodiversity hotspots for conservation priorities. Nature 403: 853–858.

Negrelle, R. R. B. 1995. Sprouting after uprooting of canopy trees in the Atlantic rainforest of Brazil. Biotropica 27: 448–454.

Neish, A. C. 1965. Coumarins, phenylpropanes and lignins, pages 581–617 in J. Bonner (editor), Plant Biochemistry. Academic Press, New York.

Nerval, G. de. 1922. Les Vers Dorés. Sant'Andrea Marcerou et Compagnie, Paris.

Ng, F. S. P. 1977. Shyness in trees. Nature Malaysiana 2(2): 34–37.

Ninio, J. 1994. Stéréomagie. Éditions du Seuil, Paris.

Nowarra, H. J. 1993. Die deutsche Luftrüstung, 1933–1945. Bernard und Graefe Verlag, Coblenz.

NRC (National Research Council). 1984. Casuarinas: Nitrogen-Fixing Trees for Adverse Sites. National Academy Press, Washington, D.C.

Oldeman, R. A. A. 1972. L'architecture de la végétation ripicole forestière des fleuves et criques Guyanais. Adansonia, nouv. sér., 12: 253–265.

Oldeman, R. A. A. 1974. L'Architecture de la Forêt Guyanaise. ORSTOM Mémoire 73, Paris.

Parenti, L. R. 1986. Bilateral asymmetry in phallostetid fishes *(Atherinomorpha)* with descriptions of a new species from Sarawak. Proceedings of the California Academy of Sciences 44: 225–236.

Passet, R. 1979. L'Économie et le Vivant. Payot, Paris.

Pasteur, G. 1995. Biologie et Mimétismes, de la Molecule à l'Homme. Nathan, Paris.

Pelt, J.-M. 1986. Mes Plus Belles Histoires de Plantes. Fayard, Paris.

Pelton, J. 1953. Studies on the life history of *Symphoricarpus occidentalis* Hook. in Minnesota. Ecological Monographs 23: 17–39.

Perrin, Y. 1994. Optimisation du Rejeunissement des Clones d'*Hevea brasiliensis* en Vue de Leur Microbouturage in Vitro: Mis en Oeuvre de Techniques Morphogénétiques et Biochimiques. Thèse, Université Montpellier II.

Perrot, V., S. Richards, and M. Valero. 1991. Transition from haploidy to diploidy. Nature 351: 315–317.

Pirozynski, K. A. 1988. Coevolution by horizontal gene transfer: a speculation on the role of fungi, pages 247–268 in K. A. Pirozynski and D. L. Hawksworth (editors), Coevolution of Fungi with Plants and Animals. Academic Press, San Diego.

Polak, M., and R. Trivers. 1994. The science of symmetry in biology. Trends in Ecology and Evolution 9(4): 122–124.

Policansky, D. 1982. The asymmetry of flounders. Scientific American 246(5): 116–122.

Pollan, M. 2001. The Botany of Desire: a Plant's-Eye View of the World. Random House, New York.

Ponge, F. 1942. Le Parti Pris des Choses. Gallimard, Paris.

Prochiantz, A. 1988. Les Strategies de l'Embryon: Embryons, Gènes, Évolution. Presses Universitaires de France, Paris.

Prost, M. 1992. The world of plants and animals: differences and resemblances [in Polish]. Medycyna Weterynaryjna 48: 387–390.

Putz, F. E. 1994. Vines in treetops: consequences of mechanical dependence, pages 311–323 in M. D. Lowman and N. Nadkarni (editors), Forest Canopies. Academic Press, Orlando, Florida.

Putz, F. E., and N. M. Holbrook. 1986. Notes on the natural history of hemiepiphytes. Selbyana 9: 61–69.

Raper, K. B. 1940. Pseudoplasmodium formation and organization in *Dictyostelium discoideum*. Journal of the Elisha Mitchell Society 56: 241–282.

Rasmont, R. 1979. Les éponges: des métazoaires et des sociétés des cellules, pages 21–29 in C. Levy and N. Boury-Esnault (editors), Biologie des Spongaires. Colloques International de CNRS 291, Paris.

Raup, D. M. 1991. Extinction: Bad Genes or Bad Luck? W. W. Norton, New York.

Raven, J. A., D. I. Walker, K. R. Jensen, L. L. Handley, C. M. Scrimgeour, and S. G. McInroy. 2001. What fraction of the organic carbon in sacoglossans is obtained from photosynthesis by kleptoplastids? An investigation using the natural abundance of stable carbon isotopes. Marine Biology 138: 537–545.

Raynal, A. 1987. Le monde des plantes, un monde d'odeurs, in Parfums des Plantes. Museum National d'Histoire Naturelle, Paris.

Rayner, A. D. M., and N. R. Franks. 1987. Evolutionary and ecological parallels between ants and fungi. Trends in Ecology and Evolution 2: 127–133.

Reeves, H., and J. Obrenovitch. 1992. Compagnons de Voyage. Éditions du Seuil, Paris.

Reichholf, J.-H. 1993. L'Émancipation de la Vie. Flammarion, Paris.

Revardel, J.-L. 1993. Constance et Fantaisie du Vivant: Biologie et Évolution. Albin Michel, Paris.

Rey, A., et al. 1992. Dictionnaire Historique de la Langue Française. Dictionnaires Le Robert, Paris.

Richards, P. W. 1952. The Tropical Rain Forest: an Ecological Study. Cambridge University Press, Cambridge.

Rilke, R. M. 1923. The Duino Elegies, translated by David Osvald. Daimon Verlag, Einsiedeln, Switzerland.

Robert, D., and J.-C. Roland. 1989. Biologie Végétale: Characteristiques et Strategie Évolutive des Plantes, vol. 1, Organisation Cellulaire. Doin, Paris.

Roberts, E., and E. Amidon. 1999. Prayers for a Thousand Years. HarperCollins, San Francisco.

Roland, J.-C., and F. Roland. 1995. Atlas de Biologie Végétale, éd. 6. Masson, Paris.

Ronsard, P. de. 1972. Sonnets pour Helene, translated by Humberte Wolfe. Hyperion Press, Westport, Connecticut.

Rosnay, J. 1988. L'Aventure du Vivant. Éditions du Seuil, Paris.

Rudaux, L., and G. de Vaucouleurs. 1948. Astronomie: les Asters, l'Univers. Larousse, Paris.

Ruelle, D. 1991. Hasard et Chaos. Odile Jacob, Paris.

Rumelhart, M. 2000. À la Conquête de l'Infiniment Ligneux. Actes Sud, Arles.

Rumpho, M. E., E. J. Summer, and J. R. Manhart. 2000. Solar-powered sea slugs: mollusc/algal chloroplast symbiosis. Plant Physiology 123: 29–38.

Sachs, T. 1978. Patterned differentiation in plants. Differentiation 11: 65–73.

Sachs, T. 1988. Epigenetic selection: an alternative mechanism of pattern formation. Journal of Theoretical Biology 134: 547–559.

Sachs, T. 1994. Variable development as a basis for variable pattern formation. Journal of Theoretical Biology 170: 423–425.

Sachs, T., A. Novoplansky, and D. Cohen. 1993. Plants as competing populations of redundant organs. Plant Cell and Environment 16: 765–770.

Saint-Exupéry, A. de. 1943. Le Petit Prince. (The Little Prince, translated by K. Woods.) Harcourt, Brace and Company, New York.

Santiago Caicedo, D. 1993. El mundo y sus cosas, pages i–iii in P. Levya (editor), Colombia Pacifico, vol. 2. Fondo FEN, Santa Fé de Bogotá.

Sattler, R. 1988. Homeosis in plants. American Journal of Botany 75: 1606–1617.

Sattler, R., and B. Jeune. 1992. Multivariate analysis confirms the continuum view of plant form. Annals of Botany 69: 249–262.

Schmidt, R. J., B. Veit, M. A. Mandel, S. Hakes, and M. F. Yanofsky. 1993. Identification and molecular characterization of ZAGI, the maize homologue of the Arabidopsis homeotic gene AGAMOUS. Plant Cell 5: 729–737.

Schonen, S. de. 1989. Symétrie et cerveau, pages 143–159 in E. Noël and G. Minot (editors), La Symétrie au Jour d'Hui. Éditions du Seuil, Paris.

Schumacher, H. 1977. L'Univers Inconnu des Coreaux. Elsevier Séquoia, Paris.

Sellato, B. 1989. Hornbill and Dragon—Kalimantan-Sarawak-Sabah-Brunei. Elf-Aquitaine Indonesia, Jakarta.

Sharp, L. W. 1926. An Introduction to Cytology. McGraw-Hill, New York.

Shepard, P. 1996. Traces of an Omnivore. Island Press, Washington, D.C.

Shigo, A. L. 1991. A New Tree Biology. Shigo and Trees, Durham, New Hampshire.

Silander, J. A. 1985. Microevolution in clonal plants, pages 107–152 in J. B. C. Jackson, L. W. Buss, and R. E. Cook (editors), Population Biology and Evolution of Clonal Organisms. Yale University Press, New Haven.

Simpson, G. G. 1950. The Meaning of Evolution: a Study of the History of Life and of Its Significance for Man. Yale University Press, New Haven.

Slatkin, M. 1987. Somatic mutations as an evolutionary force, pages 19–30 in P. J. Greenwood, P. H. Harvey, and M. Slatkin (editors), Evolution: Essays in Honour of J. Maynard Smith. Cambridge University Press, Cambridge.

Soltis, P. S., D. E. Soltis, and M. W. Chase. 1999. Angiosperm phylogeny inferred from multiple genes as a tool for comparative biology. Nature 402: 402–404.

Souèges, R. 1919. Les premières divisions de l'oeuf et les différentiations du suspenseur chez le Capsella bursa-pastoris Moench. Annales des Sciences Botaniques, sér. 10, 1: 1–28.

Souèges, R. 1935. La Segmentation. Hermann et Compagnie, Paris.

Southwood, T. R. E. 1985. Interactions of plants and animals: patterns and processes. Oikos 44: 5–11.

Starling, E. 1905. The Chemical Correlation of the Functions of the Body. Cronian Lecture (20 June), Royal College of Physicians, London.

Stephenson, T. A. 1931. Development and formation of colonies of Pocillopora and Porites. Scientific Reports of the Great Barrier Expedition 3: 113–134.

Stevens, P. F. 1984. Metaphors and typology in the development of botanical systematics, 1690–1960, or the art of putting old wine into new bottles. Taxon 33: 169–211.

Stevens, P. S. 1974. Patterns in Nature. Little, Brown and Company, Boston.

Supervielle, J. 1925 (éd. 4, 1937). Gravitations, Poèmes. Éditions Gallimard, Paris.

Sutherland, W. J., and A. R. Watkinson. 1986. Do plants evolve differently? Nature 320: 305.

Tardi, J. 1978. Monies en Folie. Casterman, Paris.

Taylor, T. N., and E. L. Taylor. 1993. The Biology and Evolution of Fossil Plants. Prentice-Hall, Englewood Cliffs, New Jersey.

Thom, R. 1972. Stabilité Structurelle et Morphogénèse: Essai d'une Théorie Générale des Modèles. (1975. Structural Stability and Morphogenesis: an Outline of a General Theory of Models, translated by D. H. Fowler, as updated by the author.) W. A. Benjamin, Reading, Massachusetts.

Thom, R. 1988. Esquisse d'une Sémiophysique. InterÉditions, Paris.

Thom, R. 1993. Comparison des morphogenèses animale et végétale, pages 137–138 in D. Barabe and R. Brunet (editors), Morphogenèse et Dynamique. Orbis Publishing, London.

Thompson, J. D., E. A. Herre, J. L. Hamrick, and J. L. Astone. 1991. Genetic mosaics in strangler figs: implications for tropical conservation. Science 254: 1214–1216.

Till-Bottraud, I., and P. H. Gouyon. 1992. Intra- versus interplant Batesian mimicry? A model on cyanogenesis and herbivory in clonal plants. American Naturalist 139: 509–520.

Tonegawa, S. 1983. Somatic generation of antibody diversity. Nature 302: 575–581.

Trewavas, A. J. 1981. How do growth substances work? Plant, Cell and Environment 4: 203–228.

Trewavas, A. J. 1982. Possible control points in plant development, pages 7–27 in H. Smith and D. Grierson (editors), The Molecular Biology of Plant Development. University of California Press, Berkeley.

Trewavas, A. J. 1986. Resource allocation under poor growth conditions: a major role for growth substances in developmental plasticity, pages 31–76 in D. H. Jennings and A. J. Trewavas (editors), Plasticity in Plants. Cambridge University Press, Cambridge.

Valantin, M. 1996. Tel Est Pris Qui Croyait Prendre. Chai du Terral, Empreintes de Pan, Montpellier.

Valentine, J. 1978. The evolution of multicellular plants and animals. Scientific American 239(3): 140–158.

Valéry, P. 1905–07. Cahiers VIII.

Valéry, P. 1950. Histoires Brisées. Gallimard, Paris.

Valéry, P. 1956. Dialogues, translated by W. McC. Stewart. Pantheon, New York.

Van de Vyver, G., and P. Willenz. 1975. An experimental study of the life cycle of the freshwater sponge *Ephydatia fluviatalis*. Wilhelm Roux's Archive 177: 41–52.

Van Duyl, F. C., R. P. M. Bak, and J. Sytsma. 1981. The ecology of the tropical compound ascidian *Trididemnum solidum* 1. Reproductive strategy and larval behavior. Marine Ecology Progress Series 6: 35–42.

Van Hoven, W. 1991. Mortalities in kudu populations related to chemical defenses in trees, pages 535–538 in C. Edelin (editor), L'Arbre. Biologie et Developpement. Naturalia Monspeliensia, hors sér.

Van Steenis, C. G. G. J. 1976. Autonomous evolution in plants: differences in plant and animal evolution. Gardens Bulletin (Singapore) 29: 103–126.

Verhey, S. D., and T. L. Lomax. 1993. Signal transduction in vascular plants. Journal of Plant Growth Regulators 12: 179–195.

Veron, J. E. N. 1995. Corals in Space and Time: the Biology and Evolution of the Scleractinia. University of New South Wales Press, Sydney.

Vieljeux, D. 1982. L'Arbre de Rêve: l'Imaginaire dans le Test de l'Arbre. M. A. Éditions, Paris.

Vogel, S. 1988. Life's Devices: the Physical World of Animals and Plants. Princeton University Press, Princeton.

Walbot, V. 1985. On the life strategies of plants and animals. Trends in Genetics 1: 165–169.

Walbot, V. 1996. Sources and consequences of phenotypic plasticity in flowering plants. Trends in Plant Science 1: 27–32.

Walbot, V., and C. A. Cullis. 1983. The plasticity of the plant genome: Is it a requirement for success? Plant Molecular Biology Reporter 1(4): 3–11.

Walbot, V., and C. A. Cullis. 1985. Rapid genomic change in higher plants. Annual Review of Ecology and Systematics 36: 367–397.

Walker, J. R. L. 1975. The Biology of Plant Phenolics. Edward Arnold, London.

Waller, D. M. 1988. Plant morphology and reproduction, pages 203–227 in J. and L. Lovett-Doust (editors), Plant Reproductive Ecology: Patterns and Strategies. Oxford University Press, New York.

Watkinson, A. R., and J. White. 1986. Some life-history consequences of modular construction in plants. Philosophical Transactions of the Royal Society of London, B, 313: 31–51.

Weaver, J. E. 1919. The ecological relations of roots. Carnegie Institution of Washington Publication 286: 1–128.

Weismann, A. 1892. Das Keimplasma: eine Theorie der Vererbung. (1893. The Germ-plasm: a Theory of Heredity, translated by W. Newton Parker and Harriet Rönnfeldt. Scribners, New York.)

Went, F. 1971. Parallel evolution. Taxon 20: 197–225.

Went, F., and K. V. Thimann. 1937. Phytohormones. Macmillan, New York.

Weyers, J. 1984. Do plants really have hormones? New Scientist 102(1410): 9–13.

Wheeler, W. M. 1911. The ant-colony as an organism. Journal of Morphology 22: 307–325.

White, J. 1979. The plant as a metapopulation. Annual Reviews of Ecology and Systematics 10: 109–145.

White, J. 1984. Plant metamerism, pages 15–47 in R. Dirzo and J. Sarukhan (editors), Perspectives on Plant Ecology. Sinauer, Sunderland, Massachusetts.

Whitham, T. G., and C. N. Slobodchikoff. 1981. Evolution by individuals, plant-herbivore interactions and mosaics of genetic variability: the adaptive significance of somatic mutations in plants. Oecologia 49: 287–292.

Wiens, D. 1978. Mimicry in plants. Evolutionary Biology 11: 365–403.

Wilkins, M. B. 1995. Are plants intelligent?, pages 119–133 in P. Day and C. Catlow (editors), Bicycling to Utopia. Oxford University Press, Oxford.

Windle, P. 1992. The ecology of grief. BioScience 42: 363–366.

Winter, D. B., and P. J. Gearhart. 1995. Another piece in the hypermutation puzzle. Current Biology 5: 1345–1346.

Wyndham, J. 1954. The Day of the Triffids. Penguin Books, London.

Zürcher, E., M. G. Cantiani, F. Sorbetti-Guerri, and D. Michel. 1998. Tree stem diameters fluctuate with tides. Nature 392: 665–666.

Index

Lightning Source UK Ltd.
Milton Keynes UK
UKHW01f1046010518
321931UK00005B/872/P